MAPS, MYTHS & PARADIGMS

WITH A SIDE OF COPHEE

Doug Fisher

Published by Cophee House Books 2018.
First Edition

Maps, Myths & Paradigms: With a Side of COPHEE /
Doug Fisher

ISBN: 978-0-578-40870-5
Library of Congress Control Number: 2018912899

Includes Index.
1. Ancient Maps. 2. Lost civilizations. 3. Plate tectonics. 4. Expanding Earth.
I. Title

To my wonderful wife.
I can never thank you enough for your support and
your infinite patience.

Contents

PART I

ANCIENT MAPS

Myths and Mysteries

CHAPTER 1

HERE THERE BE DRAGONS

I grew up in a family of modest means in the 1970s. When I was young, my parents made a seemingly innocuous purchase that was greeted with much curiosity and excitement: a world atlas. It was a welcome luxury that opened my eyes to a world of wonder; I recall poring over it for countless hours studying country and state boundaries, capitals, mountain ranges, oceans, seas, and various waterways.

It was not just the geographic details pertaining to different regions of the world that captured my attention. I was also in awe of the level of artistic design, the textures and patterns, which went into the maps to distinguish different terrains from complex mountain ranges to low-lying plains and foothills. Having an insatiable curiosity and limited entertainment options, I put an inordinate amount of miles on that hardbound treasure. Those early years kindled a fascination with maps in me that remains to this day.

My interest was further fueled by the fantastic tales of pirates and treasure maps I found in an assortment of books and movies throughout my youth. The discovery of an old tattered map inscribed with an "X" usually marked an escalation of mystery and adventure en route to the treasure that lay ahead.

For our ancestors of a few centuries past, adventure and treasure lay in yet-to-be-discovered sea routes. At the turn of the sixteenth

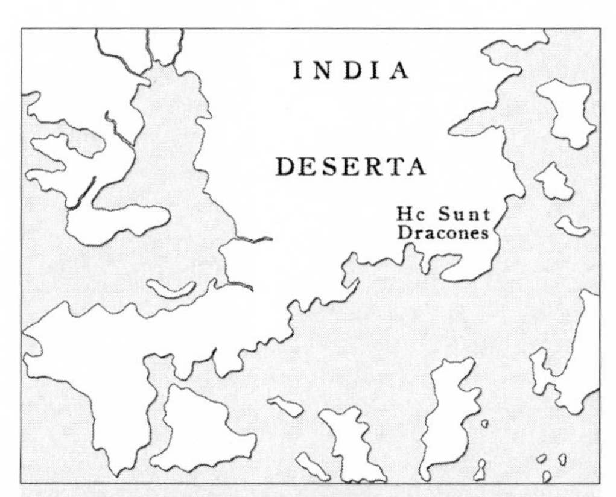

Figure 1. *A section of the Lenox globe c. 1510 upon which we find the Latin inscription,* "Hc Sunt Dracones," *which translates into English,* "*Here be dragons.*"

century, European explorers sought adventure and fortune in the pursuit of alternative routes to the spice-rich West Indies. But as many a ship's log can attest, these journeys were not without risk. Maps of the day conveyed the risks mapmakers feared lay in the uncharted waters and far-off lands. One globe from this period, the Lenox Globe (Fig. 1), contains a unique inscription cautioning the aspiring adventurer: "Here be Dragons." The inscription lies near the eastern coast of Asia and is similar to other inscriptions found on ancient Roman and medieval maps. The warnings signified uncertainties and potential dangers in regions that were not well known at the time of the various maps' charting.

Many of these maps also carried depictions of these dragons, as well as other mythical creatures believed to inhabit remote uncharted regions of the world. A medieval map known as the Hereford map depicts dragons in the mountainous terrain of India. Borrowing heavily from ancient beliefs, it is also emblazoned with depictions of bizarre creatures such as the *bonnacon*, a beast similar to a buffalo but armed with the ability to destroy would-be pursuers within a three-acre range by discharging flaming excrement. These fantastical images undoubtedly struck terror into the hearts of ancient explorers.

Another fear-inducing creature found on many old maps is the basilisk, which was considered king of the serpents and had reptilian features but the head of a chicken. It was believed to hatch from a cock's egg that had been incubated by a serpent, and both its gaze and its breath were supposedly lethal. We can take comfort in the fact that the basilisk is nearly as rare as a cock's egg.

There are other mythical creatures depicted on these maps that we

Figure 2. *A portion of the 15th century Nordic map Carta Marina, Map of the Sea, which depicts a diverse collection of ship-menacing creatures inhabiting the North Atlantic.*

would recognize today. These include land dwellers like the griffin, a creature with the body of a lion and the head and wings of an eagle, the centaur and Minotaur, half-human creatures that were also half horse and half bull respectively, and sea creatures like the ever-popular mermaid. The seas remained a great mystery to early explorers and were believed to contain all manner of sea monsters. These were drawn at tremendous scale and believed capable of enveloping and capsizing a seagoing vessel. These mythical creatures, which allegedly lurked in the unseen depths of the sea, were likely based on reported sightings of whales and giant squid, or just imaginative tales spun by sailors having returned from long voyages at sea.

These creatures would eventually disappear from maps as exploration expanded and knowledge and reason began to supplant superstition and fear. But some creatures still live on the fringes of society's imagination—and though they no longer grace conventional maps, one can find maps detailing the location of alleged sightings.

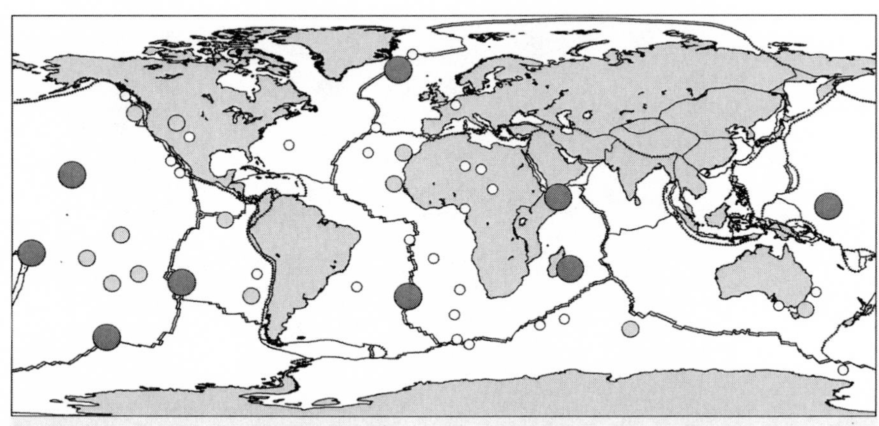

Figure 3. *Modern day dragons? Map of the world incorporating locations of perceived hotspot mantle plumes of various sizes and intensities, invisible forces believed to be lying deep beneath the planet's surface.*

These creatures include Sasquatch, often sighted in the forests of North America, as well as a lake monster lurking in the depths of Scotland's Loch Ness. Neither creature has ever been captured convincingly on film, nor have any carcasses ever been found, making the existence of these creatures highly suspect.

Yet there is global consensus that unseen forces dwell in a yet-to-be-explored frontier: the depths of Earth's interior. The purported existence of modern-day subterranean dragons has been charted throughout the globe (Fig. 3). These dragons, which take the form of tectonic forces known as hotspots or mantle plumes, are believed to spew channels of superheated magma toward Earth's surface, carving up the thin layer of seafloor crust and leaving behind scars in the form of magmatic ridges. Another invisible force believed to exist within Earth's interior is alleged to be responsible for moving tectonic plates over these hotspots. These powerful invisible forces are circulating patterns of magma referred to as convection cells.

Scientists formulated the theory of hotspots and mantle plumes to explain the existence of geological features and activities occurring on the planet's surface, specifically the creation of seafloor ridges. Perhaps the most notable of these formations is the Hawaiian-Emperor seamount chain in the Pacific. Convection cells, operating in a manner similar to conveyer belts, are believed to be transporting the Pacific seafloor plate in a northwesterly direction over the Hawaiian hotspot. This extends the length of the ridge ever southeastward as the stationary hotspot, which currently resides beneath the Hawaiian Islands, cuts into the overriding plate.

Is it possible that today's society is repeating the mistakes of the past? Could these mysterious hotspots and convection cells which lurk at unknown depths beneath the planet's surface be our modern-day dragons—powerful mythical creations born of imagination and destined to be confined there? Although their existence is explained by more logical methods than those dragons of yore, they remain similarly unseen. Yes, perhaps modern science recognizes that hotspots are merely part of a working hypothesis to explain phenomena we do not completely understand, but the reality is that many perceive their existence as fact. This work has the potential to effectively and convincingly slay these modern-day dragons and perhaps even draw scientific consensus around a new theory of Earth dynamics.

This entire work is in many respects a treasure hunt, where maps from many ages and of various designs form a trail of clues leading to a string of discoveries. The following chapters will take us to the fringe of science and history to confront dragons put in place by men of the past and others established by the best scientific minds of our time. So take map and compass in hand and join me as we revisit ancient maps of Antarctica revealing startling new evidence supporting their authenticity. Witness the discovery of a 2,000-year-old map on the bottom of one world and track the location of the world's largest impact crater atop another. And finally, chart the rise of Atlantis from the depths of myth and legend and the fall of a modern paradigm from the heights of scientific consensus.

CHAPTER 2

DOWN THE RABBIT HOLE

It all began 15 years ago. I was still a skeptic when it came to many of the fringe subjects that follow. I was either oblivious to them or, at most, considered them interesting curiosities. At the time, I was more curious about contemporary issues such as climate change. This is why my attention was drawn to a satellite image released by NASA in the spring of 2002 which displayed a sizable portion of the Larsen B Ice Shelf breaking free from Antarctica. It was an extraordinary image; the section of ice that was threatening to collapse was nearly the size of the state of New Hampshire. Scientists claimed that it was a casualty of the ongoing effects of global warming.

Intrigued, I began to research the Larsen B Ice Shelf. I performed a web search for maps of Antarctica, scrolling through images of maps that were a mix of computer and hand-drafted designs. It was not long before I found myself staring at an unusual map among the results. A strong sense of curiosity coupled with a fascination with maps made it impossible for me to ignore.

The map did not seem to share any similarities with present-day Antarctica; it was clearly out of place but its design was intriguing. Little did I know this map would send me down a virtual rabbit hole that would blur the lines between myth and reality. I would find myself exiting the other side having gone full circle, revisiting the subject of

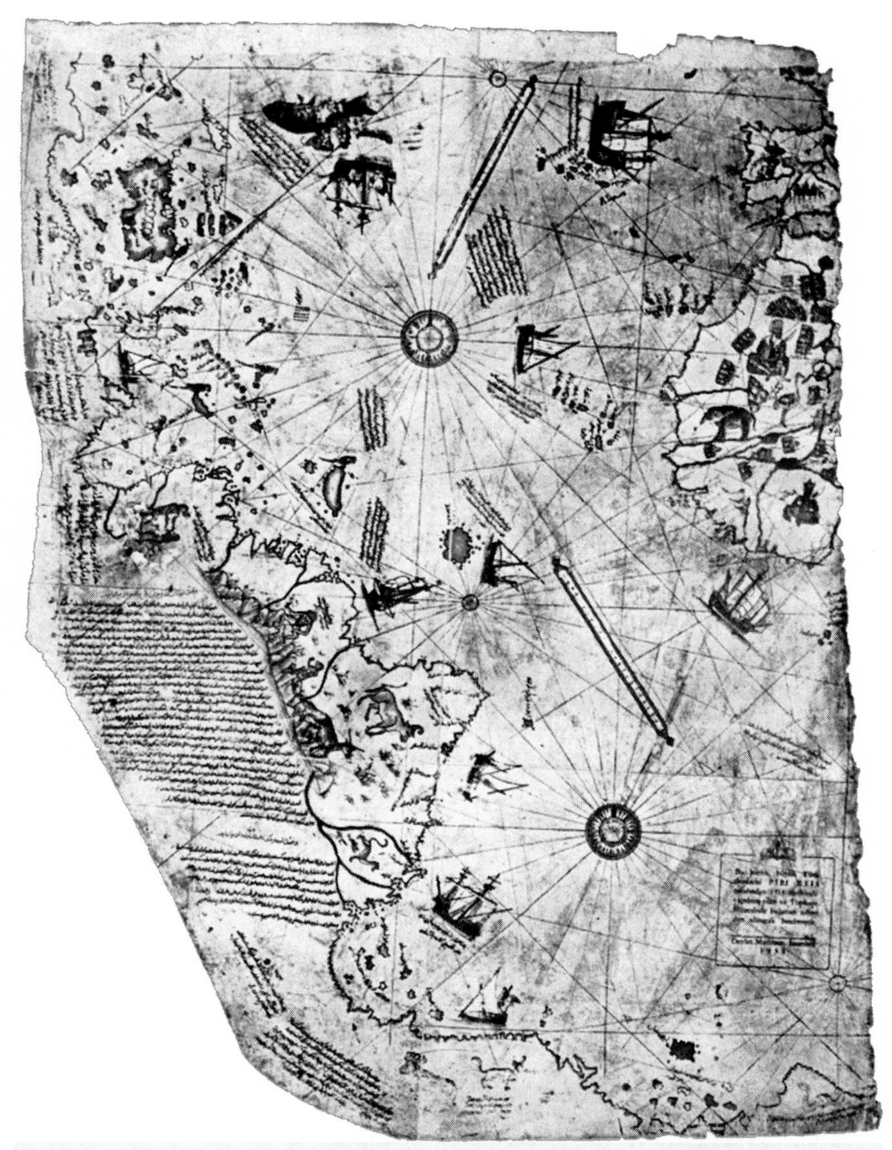

Figure 4. *The 1513 Piri Re'is World Map, of which only the lower left quarter remains and is displayed in this image. Charles Hapgood popularized the map in his book "Maps of the Ancient Sea Kings," determining that the map maintained an accurate portrayal of the Antarctic continent some 300 years prior to its known discovery.*

global warming and revealing newly realized devastating effects linked to the warming process.

The image I had stumbled across was a map produced in the early sixteenth century by Ottoman Admiral and cartographer Piri Re'is (Fig. 4). The map is dated 1513 and has survived only in part; only the lower left quarter of the parchment remains. The map is portolan in design, a common navigational map of that period that utilizes rhumb lines, lines that extend out from each of the 32 points on a compass. An array of compasses laid out on the map along with their associated rhumb lines allow the navigator to gauge headings between ports when plotting a ship's course.

The Piri Re'is map first gained fame back in 1929 when it was discovered in the Library of the Topkapı Palace in Istanbul, Turkey. Inscribed on the map is a note from Piri Re'is stating that in the making of his map he used "a map of the western regions drawn by Colombo (Columbus)." At the time of its discovery, this map's main claim to fame was that it was the only map derived directly from a long-lost map known to have been produced by Columbus—a map that charted his discovery of the West Indies.

It gained much greater notoriety in the 1960s through the studies of Charles Hapgood, a professor of history and anthropology. When Hapgood became aware of the map, he and his students at Keene State College began investigating it and became convinced that it depicted Antarctica along its lower border. This was an astonishing hypothesis because the map dates to some 300 years prior to Antarctica's known discovery. Hapgood would later go on to document and publish their findings in the book *Maps of the Ancient Sea Kings*.

At first glance, the map's portrayal of South America is unique. The coast lines the lower left portion of the map and extends eastward out across the bottom edge. But at the point where the land transitions laterally, Hapgood identifies a prominent point of land as the Palmer Peninsula; he argues that Antarctica extends eastward from there (Fig. 5).

It is an interesting theory, but one fraught with issues. One of the many problems with Hapgood's proposal is the omission of a 600-mile stretch of sea lying between Palmer and the southern tip of South

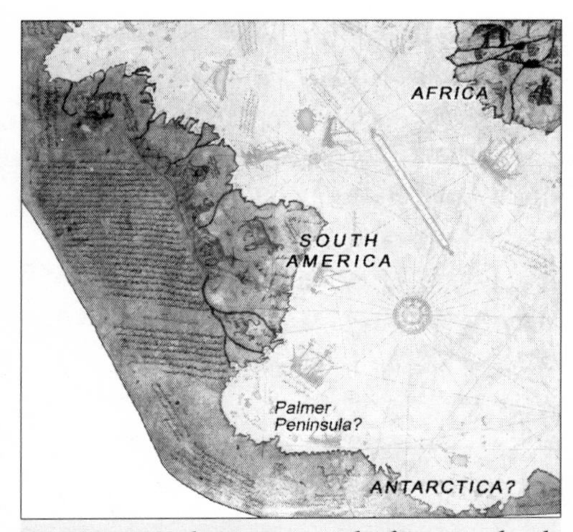

Figure 5. *South America and Africa are clearly represented in the Piri Re'is map. What is not so clear is Hapgood's claim that Antarctica and the Palmer Peninsula are also represented. Note that the map makes no effort to distinguish a separate Antarctic continent. It merely overextends the South American continent.*

America. This significant span of water, now known as the Drake Passage, is omitted entirely from the Piri Re'is map. Not even as much as a small rivulet exists to show the separation between two continents.

Not only does Hapgood acknowledge this omission of the Drake Passage; he also accepts that 1,500 miles of coastline, extending from Rio de Janiero in the north to Bahia Blanco in the south, has been omitted from the map. He accepts these major omissions and moves forward with his hypothesis anyway, assuming that errors in the map had amassed over time through multiple chartings. But is this the best and most accurate approach to analyzing the Piri Re'is map? If we are basing our analysis on a fragment of a larger map, is writing off thousands of miles of sea and coastline a practical method for making an argument in favor of the map's highly unlikely inclusion of Antarctica?

A more practical approach would be to compare the map to other maps from the same period. There is a progression in the history of cartographic design; for any given historical period, certain designs and components are unique and specific to that period. This is, of course, due to newer designs and improvements being shared constantly among cartographers.

Fortunately, there happens to be another map from the same period still in existence that allows better analysis of the Piri Re'is map and reveals its original full form. That map is the 1519 Lopo Homem Map.

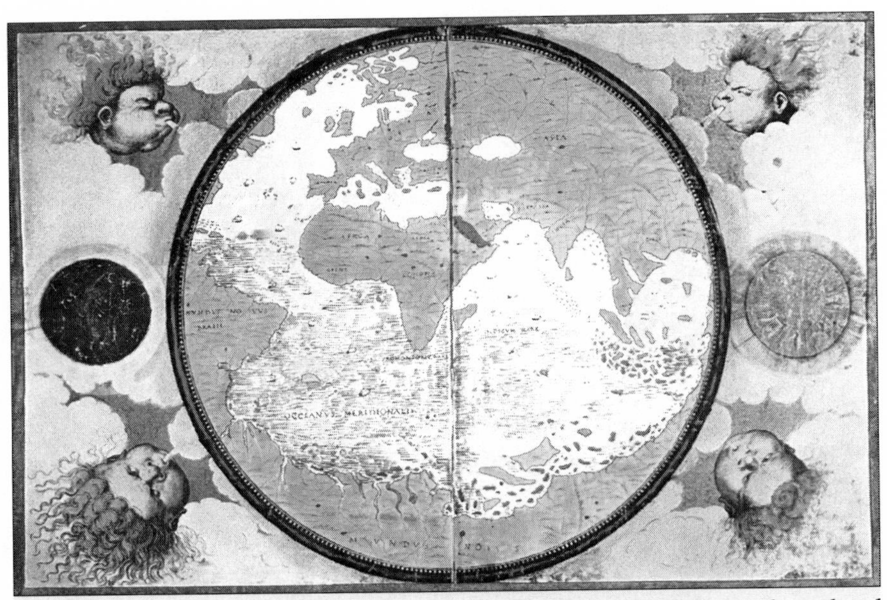

Figure 6. *The 1519 Lopo Homem Map with its depiction of a vast nearly enclosed southern sea.*

Dated only six years after the Piri Re'is map, the Lopo Homem map can be found gracing the pages of the 1519 Miller Atlas (Fig. 6).

Both maps similarly line the leftmost edge with the Americas with the majority of the western regions sheared away. The maps also share a unique portrayal of South America that is not found on any other existing map. The shared feature exists along the eastern coast: a deep narrow gulf oriented in a west-northwest direction notched into the upper end of a very large perfectly arcing coastline (Fig. 7).

More significantly, at the bottom of this arc the South American continent extends eastward, skirting the bottom of each respective map. Again, when we consider the uniqueness of these features along with the maps' closeness in dates, the odds are extremely high that much of the remaining design is similar. Where this coastline is abruptly cut off on the Piri Re'is map due to the ravages of time, the Lopo Homem fills in the void. It shows a long, unbroken coastline extending completely across the bottom of the map and rising along its eastern edge where

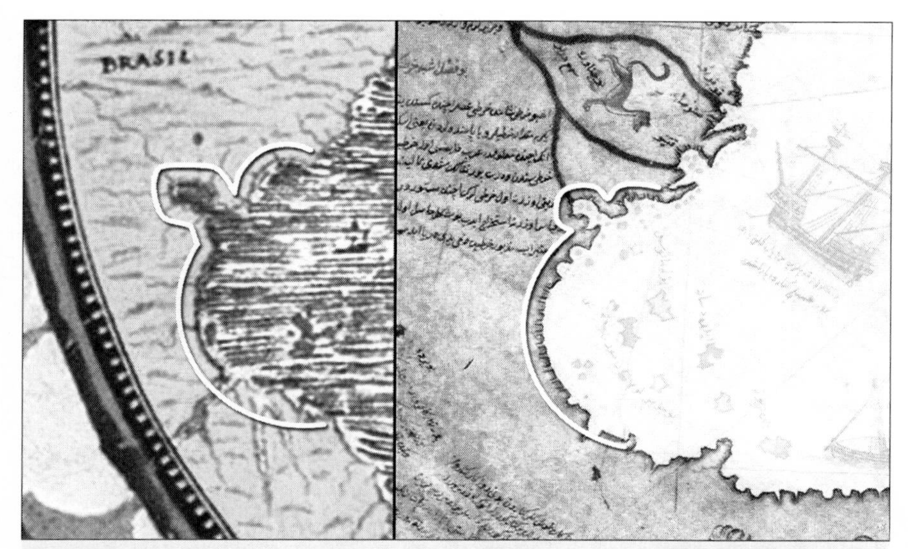

Figure 7. *Portions of the 1519 Lopo Homem (left) and 1513 Piri Re'is maps (right) sharing identical depictions of a deep gulf along the South American coast intruding on the upper end of an arcing coastline followed to the south by a lateral extension of the continent. They are features unique to these two maps dating only six years apart. Logic dictates that both maps originally shared similar depictions of an enclosed southern sea.*

it reconnects with the Asian continent. The result is a large, nearly landlocked southern sea stretching beneath the span of Africa and Asia.

An inscription on the Piri Re'is map supports the likelihood that the Piri Re'is map in its complete form once carried this same design:

> *No one now living has seen a map like this. I have composed and constructed it using about twenty maps and mappaemundi; these are the maps which were composed in the time of Alexander of the Two Horns, and which show the inhabited portion of the earth. The Arabs call these maps ja'fariya.*
>
> *I have used eight ja'fariya map, an Arab map of India and four recent Portuguese maps – these maps show the sea of Sind (Sindhu-sagara), India (Arabian Sea) and China according to mathematical principles – and also a map of the western regions drawn by Colombo (Columbus). The final form was arrived at by reducing all these maps to the same scale. Therefore the present*

> *map is as accurate for the Seven Seas as the maps of our own countries used by sailors.*

There are two keys here. The first is that the source maps used were claimed to be from the time of "Alexander of the Two Horns," a reference to Alexander the Great. This was a clear confusion of Ptolemy I Soter, one of Alexander's generals from the third century BCE, with the much later Greek scholar Claudius Ptolemy of the second century CE, a mathematician and geographer who made significant contributions to the cartographic world by applying "mathematical principles" to mapmaking.

The second key lies in his description of his process of scaling multiple source maps together into one world map. The apparent claim that some of the source maps were produced by Claudius Ptolemy is significant because Ptolemy originated the idea of a large enclosed southern sea. He provides the basic design for this southern sea in his historic work *Geographia*, in which he describes the Indian Ocean as the world's largest enclosed sea, with an unknown southern land linking the tip of Africa over to Asia:

> *We may say the same of the Indian sea, for with its gulfs, the Arabian, the Persian, the Gangetic, and that which is called the Great gulf, it is entirely shut in, like the Caspian, by land on all sides. Wherefore the entire earth consists of three continents, Asia, Africa, and Europe. Asia is joined to Africa by the part of Arabia enclosed by our sea and the Gulf of Arabia, and by the unknown land which is washed by the Indian sea ... Of the seas surrounded by land, as has been said before, the first in size is the Indian sea.*[1]

As on the Lopo Homem map, we find the same extension of Asia down and across the bottom of the Ptolemy's second-century world map, creating an "unknown land which is washed by the Indian sea" (Fig. 8). It is most certainly the same land extending out along the bottom of the Piri Re'is map.

Ptolemy's map, however, depicted a fully enclosed sea with the

---

[1] *Geography of Claudius Ptolemy*, translated by Edward Luther Stevenson, 1932, Book VII Ch. V

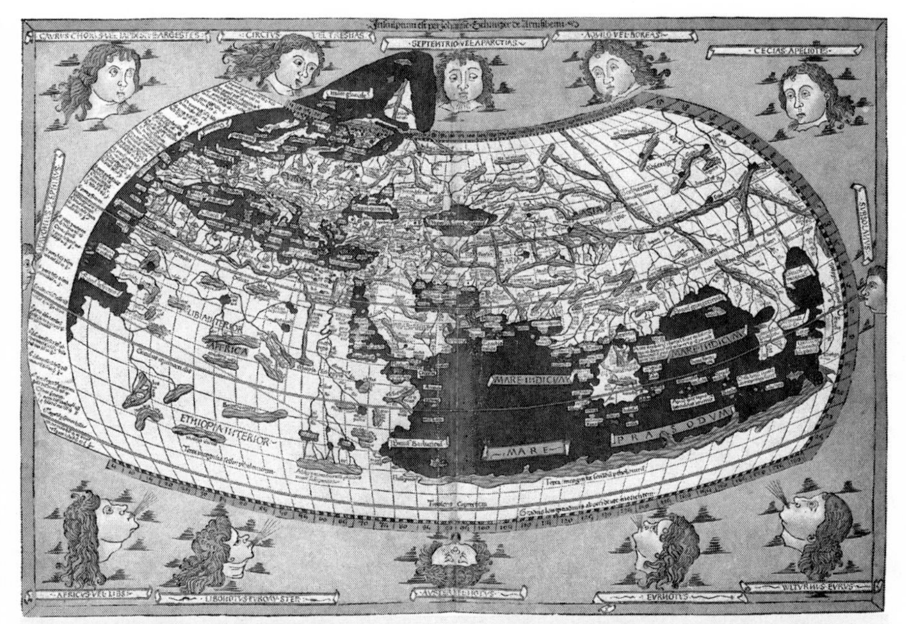

Figure 8. *Ptolemy's 2nd century world map with its depiction of a vast enclosed southern sea. (Ulm edition of Ptolemy's world map, 1482)*

African continent forming its western shore and extending down and laterally across the sea's southern perimeter. So why would Piri Re'is suddenly substitute the South American continent as the western shore?

The answer is clear: the world had evolved. Piri Re'is and other cartographers of his time were well aware that Africa did not extend out in the manner Ptolemy described. By the time of Piri Re'is, travel around the southern tip of Africa was commonplace, that being a popular trade route. The Americas had been discovered and were being explored in hopes of establishing an alternate route to the East.

The problem these early sixteenth-century cartographers were encountering was how to depict the newly discovered South American continent in its entirety when only its northern extremities had been explored and charted. How can you draw a continent whose full shape and size remain unknown? With a variety of maps at his disposal, including versions of Claudius Ptolemy's world map, Piri Re'is obviously surmised that Ptolemy's erroneous depiction of an enclosed sea might have had merit and that it had been based on a legitimate geographical concept. It appears that he adapted the concept to his map by rescaling the sea laterally and substituting South America for Africa as the sea's western shore.

Cartographers like Piri Re'is would draw upon recent maps for their design, but when these maps failed to provide details beyond currently charted regions, they would look to older maps for possible enlightenment. If the ancient concept or design seemed reasonably suited, the design was scaled, as Piri Re'is himself states, to a new map to fill in the missing gaps. Ptolemy's ancient theoretical design served Piri Re'is better than no reference at all to an unknown land.

Thus, with the aid of an inscription on his map and a little research, we are able to resolve the mystery of the Piri Re'is map. And while it is easy to refute the notion that the Piri Re'is map portrays Antarctica, Hapgood also included another sixteenth-century map in his book that would prove far more difficult to dismiss.

Looking to extend his research beyond the limits of the Piri Re'is map, Hapgood visited the Library of Congress in the winter of 1959. He had requested that fifteenth- and sixteenth-century maps carrying depictions of a southern continent be gathered for his inspection. To his surprise, the staff had exceeded all expectation having located and laid out several hundred maps for his perusal. Many long, laborious days of poring over the maps yielded some interesting finds but nothing of extreme significance. Then, one day, all Hapgood's hard work was rewarded. With the simple turn of a page, he found himself mesmerized by what lay in front of him. Hapgood was so stunned by what he saw, an ancient image depicted with such startling accuracy, he was instantly convinced he was gazing upon an authentic ancient map of Antarctica.

CHAPTER 3

ANCIENT MAPS OF ANTARCTICA

Among all the maps of an Antarctic continent produced at the turn of the sixteenth century, none are more remarkable than those by French mathematician and cartographer Oronteus Finaeus. In contrast to the Piri Re'is World Map, Finaeus' maps not only present Antarctica as an independent landmass, but also render the continent with amazing accuracy.

Finaeus' 1531 World Map (Fig. 9) is rendered on a double-cordiform projection, a design developed and popularized during the sixteenth century which places the world onto two opposed heart-shaped hemispheres, one depicting the north and the other the south. The heart-shaped layout renders the Antarctic landform with a slight distortion, with swept-back features along the Weddell Sea. But even in its distortion, it is still remarkably similar to the appearance of modern-day Antarctica on a standard polar projection. One can easily understand Hapgood's reaction of awe and amazement when he first gazed upon it. While our current view of history dictates that this cannot be an authentic map of Antarctica due to the perceived limits of ancient exploration and mapmaking, the accuracy in Finaeus' design strongly suggests otherwise.

The continent carries the Latin inscription "Terra Australis recenter inventa, sed nondum plane cognita," which translates as "Southern

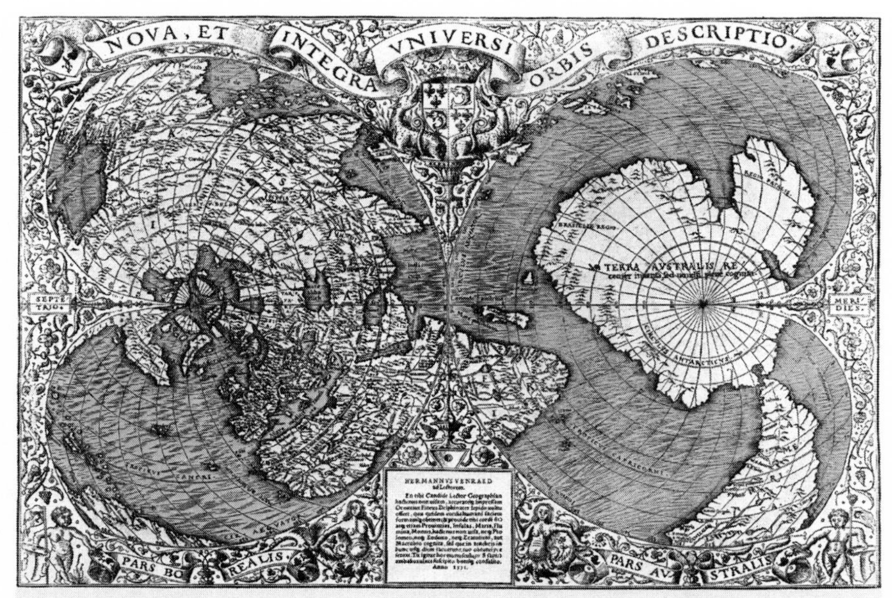

Figure 9. *Oronteus Finaeus 1531 World Map with a mysterious landmass occupying a large portion of the southern hemisphere (right). Considering the continent's location and remarkable resemblance to Antarctica, one can easily understand Hapgood's reaction upon first viewing it.*

land recently discovered but not yet fully explored." Had it been fully explored, it would have been found to be a small landform known today as Tierra del Fuego, which sits at the southern tip of Argentina. Eleven years prior to the creation of this map, Ferdinand Magellan discovered the strait linking the Atlantic to the Pacific and Finaeus was one of many attempting to provide a depiction of the unexplored land forming the strait's southern coast. Hence the massive continent positioned just a few miles off the tip of South America to allow for Magellan's strait—an erroneous depiction arrived at in much the same manner as Piri Re'is arrived at his earlier extension of South America.

Shaping a Continent

If we view Finaeus' 1531 Antarctica and a modern-day map of the continent as they would each appear laid out on a standard polar projection (Fig. 10), we can proceed with a comparative analysis that

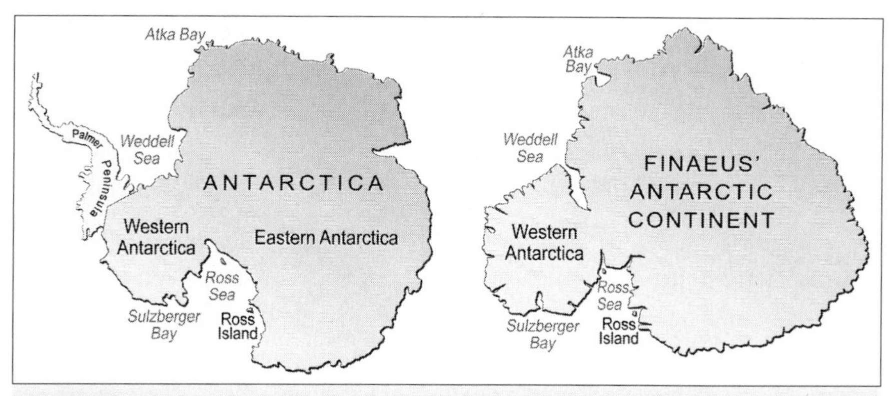

Figure 10. *Modern map of Antarctica with the Palmer Peninsula faded out (left) alongside Oronteus Finaeus' map of the continent (right), displayed as they would both appear on a standard polar projection. Both versions consist of a large, elongated eastern landmass that is roughly 1½ times taller than it is wide, called Greater or Eastern Antarctica. Protruding perpendicularly off the upper western side of this mass is a smaller and uniquely square-shaped landmass, Western Antarctica.*

more clearly reveals Finaeus' dramatic accuracy. I should first point out one of the map's more noticeable discrepancies. Finaeus omits a substantial landform extending off the northwest corner of Western Antarctica: the Palmer Peninsula. In his book *Maps of the Ancient Sea Kings*, Charles Hapgood addresses this anomaly by proposing two sites on Finaeus' map as possible representations of a scaled-down peninsula. These suggested sites are highly unlikely—not only because they in no way resemble the peninsula, but because they are also located far from where the peninsula actually extends from the continent. Rather than correctly placing the peninsula at the northwest corner of Western Antarctica, Hapgood locates it at the northeast corner and unrealistically postulates the second possible location on Eastern Antarctica. We should expect a more sound explanation for the omission, but before we tackle that issue, let us first review the rest of Finaeus' design.

Beyond the omission of the Palmer Peninsula, the continent is composed of two major landforms. Finaeus correctly depicts the main landform as a large elongated mass that is roughly 1½ times taller than it is wide. This part of Antarctica is called Greater or Eastern Antarctica. Protruding westward off this mass is the second much smaller landmass, called Lesser or Western Antarctica. Finaeus accurately aligns it with its flat northern and southern coastlines lying near perpendicular to the larger Eastern Antarctic landmass with a slight tapering away of the two coastlines toward the west. Finaeus' shape, size, placement, and orientation of Western Antarctica in relation to Eastern Antarctica are extremely impressive; and indeed, altering any one of these aspects would have greatly compromised the integrity of the overall design. The critical arrangement accurately creates a wide-angled Weddell Sea—sans Palmer Peninsula—clockwise of a deep-set Ross Sea.

As we take a closer look at Finaeus' rendering of Western Antarctica, we find some very uncanny similarities to an Antarctic continent that had yet to be discovered. Like the actual continent, Finaeus' Western Antarctica can be described as fairly squarish in proportion, bearing some relatively flat coastlines. It is astonishing that Finaeus constructs this portion of the continent in this way; the squarish multifaceted shape does not appear natural, contrasting significantly from other landforms throughout his map. If this landmass were merely an artful contrivance, it seems odd that Finaeus was compelled to go with such a seemingly unnatural shape and choose not to extend it to the rest of the continent. Even more surprising is that he happened upon a design that is also unique in the real world and limited to Western Antarctica. Finaeus accurately portrays the far western end of Western Antarctica as a very flat coastline running parallel to Eastern Antarctica. Following this coastline southward, a flat-angled chamfer occurs at the joining of this coast with Western Antarctica's southeastern coast. Here again, Finaeus continues to defy the laws of probability by accurately cutting a bay into the chamfered coastline, creating an approximate likeness of Sulzberger Bay. This is the first of four bays included on Finaeus' design, and it hits the mark for location.

Where the two landmasses converge in the south, Finaeus accurately portrays the second of four bays, this one cutting deeply into Western Antarctica's southern shore and pointing upward toward the northern convergence point along the Weddell Sea, establishing a linear divide between Eastern and Western Antarctica.[2]

At the mouth of this waterway is one of only two islands Finaeus associates with the continent. It appears to match a small island currently encased in the Ross Ice Shelf with only an ice dome indicating its presence. Moving beyond this island, the coastline forms the western coast of Victoria Land or eastern shore of the Ross Sea. This curving coastline is interrupted by an amazingly accurate set of features. Finaeus correctly depicts the Ross Sea's eastern shore as having a singular, rather pronounced point projecting into the sea, and he pairs it with his second and last island depiction corresponding to Ross Island properly set along its southern coast. The combination of the large point and Ross Island as depicted on the Finaeus map are extraordinarily accurate in both proportion and positional relationship, closely mirroring the actual land features. Ross Island is composed of three volcanoes, the largest of those being Mount Erebus, which rises to a towering height of 12,444 feet. Such a conspicuous island could have served as an important landmark and guide for ancient mariners and necessitated its inclusion on the source map.

Following the coast further around Eastern Antarctica, it is hard to equate specific features on Finaeus' map with Antarctic landforms until we make it back to the northernmost point of Eastern Antarctica. Finaeus splits this forked point with the insertion of a small bay. This matches up to Antarctica's Atka Bay located on Queen Maud Land. The level of accuracy in detailing Atka Bay can be better appreciated by contrasting the significant structural differences between it and Sulzberger Bay. Finaeus captures the essence of Sulzberger Bay as a body of water recessed into the side of a flat coastline, while accurately rendering Atka Bay as a bay nestled between two points of land extending out into the South Atlantic. We will get to the fourth and final bay shortly, but for now, Finaeus' accuracy on bay placement rests at three for three.

The fact that Finaeus does well in proportioning the eastern part

[2] It should be noted that this bay is actually drawn as the mouth of a river with the typical oversized wedge shape trailing off into a river. So it is a bit of a stretch to include it with Finaeus' four bays, but the similarity to an existing bay sandwiched between the Gould and Amundsen coasts is too significant to ignore.

of the continent but shows a lack of accurate definition of coastal features lying between Ross Island and Atka Bay may reveal the limits of the cartographer's familiarity with the continent, suggesting that his civilization most often frequented the western portion of the continent. This is a common trait shared with ancient maps. For instance, ancient Greek Maps contain highly accurate renderings of the Mediterranean, an area well-traveled by Greek sailors and documented by cartographers of the time, but their depictions of regions beyond the Mediterranean like Africa and Asia have very little resemblance to the actual continents in both shape and size. Similarly, Finaeus' portrayal of the area around Western Antarctica far exceeds the accuracy of his portrayal of most of Eastern Antarctica.

We can get a better grasp of this lopsided accuracy by placing a schematic template of Antarctica, designed to align with Western Antarctica and Ross Island, over Finaeus' map (Fig. 11). While Eastern Antarctica matches the overall size well, the coastlines rarely align or share common features. This changes dramatically at Atka Bay (A),

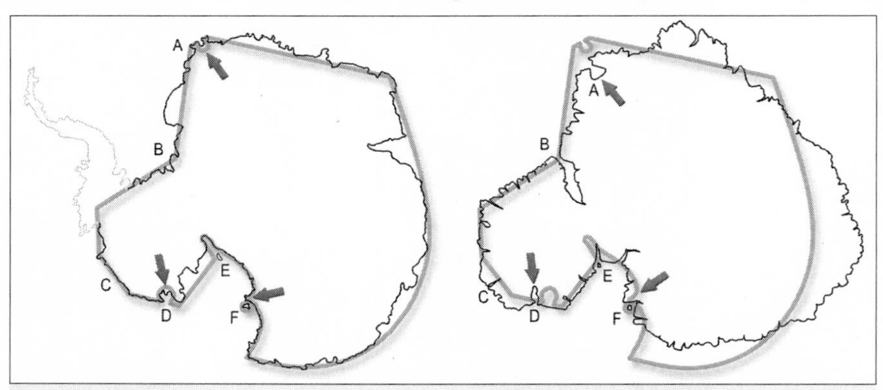

Figure 11. *Schematic template based on the shape of Antarctica overlaid onto actual Antarctica (left) and Finaeus' Antarctica (right) demonstrating the accuracy of Finaeus' design. The template is aligned to the shape of Western Antarctica and to Ross Island (F). Note that this alignment also accurately aligns Atka Bay in the north (A), and Western Antarctica along the upper half of Eastern Antarctica. Meanwhile Western Antarctica's flat westernmost coast (C) runs parallel to Eastern Antarctica, and a similarly angled chamfer extending off its southernmost point is notched by Sulzberger Bay (D), features common to both maps. Finally, Ross Island is accurately portrayed just below a lone point along the coast of Victoria Land (F).*

where we see the coast rise to a similar point and then drop into the similarly aligned Weddell Sea (B). The accuracy continues with the flat stretch of coastline at Western Antarctica's westernmost coast (C) which transitions to the similarly angled chamfer where we find Sulzberger Bay (D). Eastern and Western Antarctica converge, forming a deep bay with a lone island sitting near its mouth (E). Finally, we find Ross Island (F) positioned along the southern coast of a point extending from the coast of Victoria Land.

If Antarctica had yet to be discovered, how could such an accurate depiction exist?

Accuracy of Topography

Along with coastal formations, Finaeus provides topographic elements on his map that exhibit amazing accuracy. Again confirming the cartographer's range of geographic familiarity, the map's accuracy is mostly confined to the western region extending from Ross Island in the south to Atka Bay in the north while topographic depictions east of Atka Bay and Ross Island are strewn with extreme inaccuracies. Finaeus portrays an extensive array of nonexistent mountains lining most of Eastern Antarctica's southern and eastern coast, while along the eastern coast of the Ross Sea he accurately portrays the Queen Maud Mountain Range. He even places wider lateral mountain arrangements where the range is denser (Fig. 12). While the range is normally rendered with two laterally arranged mountains representing its standard width, five lateral mountains are drawn to show where the range extends out toward the point adjacent to Ross Island, and five more further south accurately depict the extension of the range to the southernmost point bordering the Ross Sea.

Moving directly to Finaeus' depiction of Western Antarctica, we find three well-defined mountain ranges: A) the Ellsworth Mountains, B) the Executive Committee Range, and C) the northern tip of the Queen Maud Mountains (Fig. 13). The significance of these mountain ranges on Finaeus' map is not limited to the accurate depiction of all three in the Western Antarctic region but extends to their incredibly accurate placement and orientation. Finaeus not only accurately

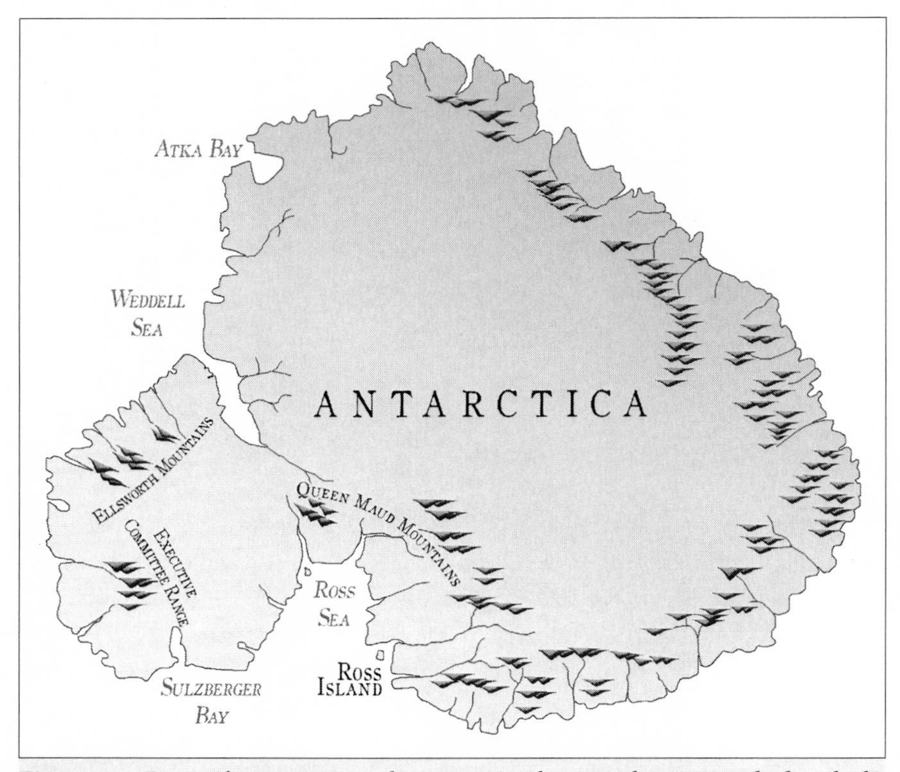

Figure 12. *Finaeus' Antarctica with mountains, bays, and seas inscribed with the names of their modern-day counterparts.*

depicts two of these as coastal ranges, but he also correctly centers the Ellsworth Mountains along the northern coast of Western Antarctica and the Executive Committee along the westernmost coast. The last mountain range he depicts, the northernmost tip of the Queen Maud Mountains, lies at the divide between Eastern and Western Antarctica; again, Finaeus accurately positions it by placing it on the eastern side of the bay. He also correctly depicts the Queen Maud Mountains as being a dividing barrier between this southern bay and a deep basin or ancient bay (D) in the north.

This is the last of the four bays depicted on Finaeus' map. It is a large waterway that he extends inland off the Weddell Sea at the junction of Western and Eastern Antarctica. This location coincides with the Foundation Ice Stream, which rests in a basin bordered on the east and

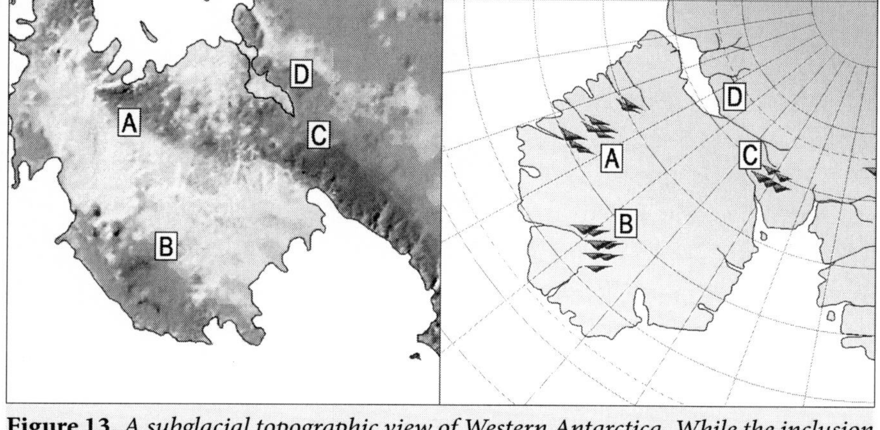

Figure 13. *A subglacial topographic view of Western Antarctica. While the inclusion of nonexistent mountain ranges along the southern and eastern coasts of Eastern Antarctica expose the cartographer's lack of full familiarity with Eastern Antarctica, the inclusion and accurate placement of A) the Ellsworth Mountains, B) the Executive Committee Range, and C) the northern tip of the Queen Maud Mountains in the western Antarctic region suggest that the ancient civilization responsible for this map frequented the western half of the continent. Meanwhile, Finaeus' incorporation of a lengthy narrow bay extending southward off the Weddell Sea mimics a basin (D) existing between two converging mountain ranges that form a similar point at its southern extremity.*

west by the Pensacola and Whitmore Mountains respectively. While the Foundation Ice Stream extends 150 miles into the basin, the basin itself extends a total length of 300 miles inland, coming to a point where the two mountain ranges converge. This virtually mirrors the deep bay depicted on Finaeus' map. This depiction of a large inland body of water placed into a tight space in a completely unaccommodating-looking mountainous region could signify a deep familiarity with the land.

Of course, the real intrigue here is that the cartographers would have had to chart this region of the continent when it was free of ice. The Foundation Ice Stream is thousands of feet thick; even Atka Bay is currently occupied by a 600-foot sheet of snow-covered ice that makes up part of the Ekstrom Ice Shelf. This presents a conflict with scientific analysis and dating of ice core samples, which have established that a deglaciated Antarctica last existed some 30 million years ago, vastly predating any civilization capable of charting the continent.

If this is an authentic map of Antarctica, we have either to believe the immensely impossible—that an advanced civilization existing more than 30 million years ago created a map that somehow endured this span of time—or the improbable, that scientific dating of the ice cap is flawed and the ice is merely thousands of years old. This adds to the list of the map's discrepancies I will address later in this work.

For now, it is important to acknowledge Finaeus' accomplishment. Not only did he create the overall shape of the Antarctic continent on his map, he accurately aligned and positioned the few bays and islands as well. It is impossible to fathom that Finaeus accomplished this degree of accuracy based on luck and guesswork alone. After all, this is not one lucky design pulled out of a collection of thousands of maps from the turn of the sixteenth century; it is one of just a handful of renditions of the continent.

Discrepancies

Focusing on the accuracy of Finaeus' map makes it nearly impossible that he could have produced it without having referenced a genuine map of the continent. Yet there are enormously glaring inaccuracies that challenge the map's accuracy. The first is the omission of the Palmer Peninsula mentioned earlier; the other is that the landmass he draws is nearly four times the actual size of Antarctica.

The missing peninsula is a significant omission. As stated earlier, I do not agree with Hapgood in associating Palmer, a substantial mass of land with an area exceeding 200,000 square miles, to a minuscule area drawn in the incorrect area and bearing no resemblance whatsoever to a peninsula.

It is far more likely that the source map omitted the peninsula entirely, deeming it a separate territory either worthy of its own map or too large to fit on the medium upon which it was drawn. Both are fairly reasonable possibilities, as this practice has occurred on other ancient maps. For instance, an ancient Roman map know as an *itinerarium* (Fig. 34, p. 62) omits most of the African continent south of a point where a lateral mountain range was believed to divide the continent in half.

Another possibility is that the ancient source map did indeed include Palmer, but when attempting to scale the source map onto a sixteenth-century map, there arose the need to alter the peninsula. This may have occurred when it was realized that the peninsula would overlap the South American continent due to Antarctica having been scaled up in size, as mentioned earlier.

In tandem with this—and since we have already broached the issue of the map depicting a deglaciated Antarctica—it is also possible that in a pre-glacial Antarctica, Palmer may have existed as an island separate from Western Antarctica. The subglacial topography certainly appears to allow for the possibility of a channel having separated Palmer Peninsula from Western Antarctica at one time where an ice bridge now connects them. As the continent froze over, the ice bridge formed, making the island appear to be a contiguous extension off the side of the continent.

Because the ancient Antarctic design was being scaled to a sixteenth-century world map, had this hypothetical Palmer Island been included on the source map, it would have been difficult for the experienced cartographer to miss that the overscaled Palmer Island overlapped a South American continent that had just recently been portrayed as an island of similar shape and size (Fig. 14). Louis Boulengier on his 1514

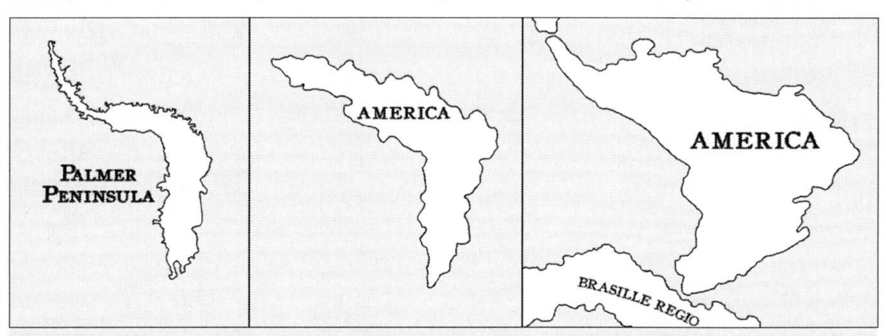

Figure 14. *Palmer Peninsula as an island (left). Had Palmer been included on Finaeus' map at his enlarged scale, it would have overlapped South America, a continent that at the turn of the sixteenth century would have shared a similar appearance. (Depictions of South America from Louis Boulengier 1514 globe and Johannes Schöner 1515 globe, center and right respectively.) Could this have led cartographers to confuse Palmer with South America believing it to be an early iteration of the continent?*

globe portrays the then little-known South American continent as a lean L-shaped island very similar in proportion to Palmer. Johannes Schöner's 1515 world globe depicts a widened version of that same South American island with an extension off the upper end, representing a portion of the Isthmus of Panama, that gives the landmass an overall S-curve appearance—also very similar to the shape of Palmer.

This overlapping of like landmasses may in turn have influenced the mapmaker to replace the source map's Palmer with South America, having assumed that Palmer was an ancient errant depiction of South America. This would have been a similar process to Piri Re'is' interpretation of Ptolemy's errant depiction of Africa, which led him to substitute South America for Ptolemy's Africa as the western shore of his version of a nearly enclosed southern ocean.

Of course, it is impossible to know for certain what actually led to the omission of the Palmer Peninsula on Finaeus' map, but these are all viable options based on the precedents I have cited. The most important point is that while the omission is a true concern, it should not negate the possibility that Schöner's design was based on a genuine map of Antarctica any more than the intentional omission of a greater portion of Africa from the Roman itinerarium, which retains all other aspects of a world map, should negate the possibility of it being a world map.

Yet even if we move beyond the omission, there remains the issue of the continent's enlarged scale. Hapgood attributes the error of overscaling to a copyist confusing the 80th parallel on the source map with the Antarctic Circle. Had Hapgood spent more time rationalizing this particular theory, he would have realized how flawed it was. If the copyist confused the 80th parallel with the Antarctic Circle—66.6° latitude—and the source map was inscribed with additional latitudinal delineations as Hapgood also suggests, this would distort the entire map. The source map would have very little resemblance to Finaeus and Mercator's rendering of the continent and in turn very little resemblance to Antarctica.

The error that Hapgood suggests would have the copyist overscaling the continent's interior by enlarging it 13-plus degrees latitude in all directions but maintaining latitudinal scaling beyond the Antarctic Circle with the aid of latitudes marked on the source map. The result would actually be a major shortening of the continent's perimetric features, similar to an artist doubling or tripling the torso of a model but maintaining the limbs at their normal size. In the case of both the cartographer and the artist there is absolutely no possibility they would overlook the fact that their resulting images in no way resembled the original subject. If we intend to validate these maps as ancient chartings of Antarctica, the overscaling of the continent requires a more reasonable explanation.

CHAPTER 4

THE MAGELLAN EFFECT

I had originally set out to resolve the issue of overscaling with a focus on three well-detailed maps containing the earliest iterations of the Antarctic design. Those maps included Finaeus' world maps of 1531 and 1534 and Mercator's world map of 1538.

I was convinced that there had to exist two points used in the scaling of the continent to the map. Scaling between points is a relatively basic concept. Piri Re'is offers one example. His incorporation of Ptolemy's enclosed Indian Ocean required him to scale the body of water from its previous position and size between the two points of Africa and Asia to two points comprising South America and Asia. This is an instance of non-uniform scaling, the resizing of an object in only one direction, which subsequently alters the shape of the original object. For example, scaling a square laterally while retaining its original height distorts the square and it becomes rectangular.

Scaling an object between two points can also be done without distortion by simultaneously stretching the object equally in all directions. In this instance, the square is not only enlarged laterally but in all other directions at the same scale factor retaining the square's original appearance though much larger. This is the type of scaling suspected of Finaeus' southern continent, which overlays and conforms

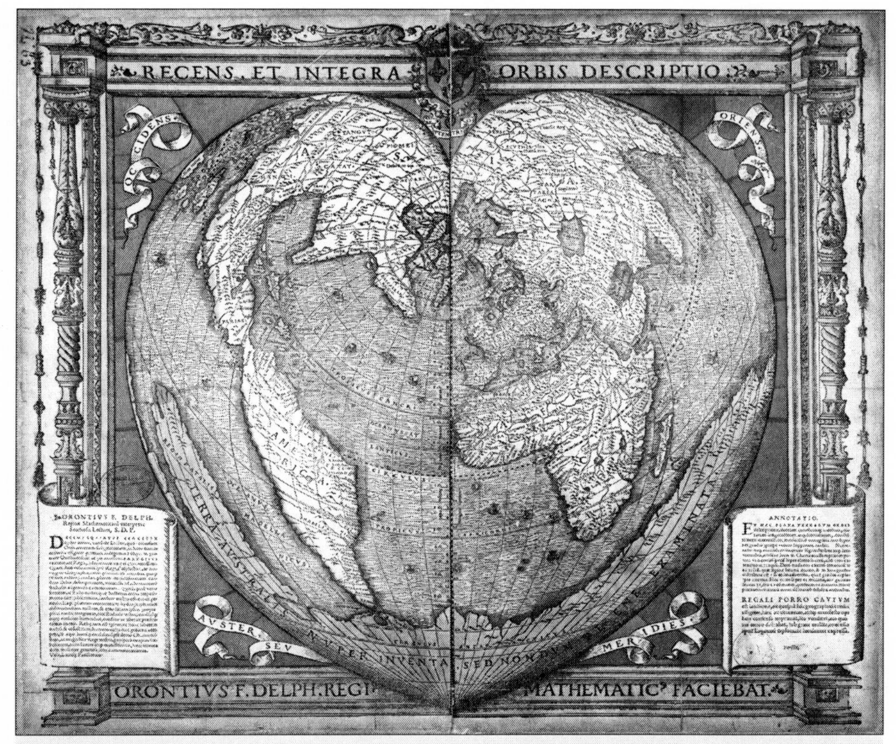

Figure 15. *Finaeus' mysterious southern continent gracing the lower perimeter of his 1534 cordiform (heart-shaped projection) map.*

quite well to Antarctica despite Finaeus' design being drawn several times larger.

Yet I could not seem to determine both scaling points. Identification of the first scaling point seemed rather obvious, as it was linked to a major discovery of the time. In his search for a passage to the Pacific, famed explorer Ferdinand Magellan had spent months exploring various waterways along the South American coast. He entered each potential throughway with hope and high expectation and exited with bitter disappointment—until October 21, 1520, when he finally entered the one waterway that would indelibly etch his name in the annals of history. Yet midway into his strait, Magellan found the waterway breaking off in two directions. Antonio Pigafetta, an Italian scholar who accompanied Magellan and maintained the most extensive account of the voyage, states that the mouth of one waterway lay toward the southwest and the other to the southeast (Fig. 16). Magellan sent two ships, the San Antonio and the Concepción, into the southeast waterway composed of Inútil Bay and Canal Whiteside. Only the Concepción returned, bearing the news that this was a closed waterway that did not offer passage to the Pacific.

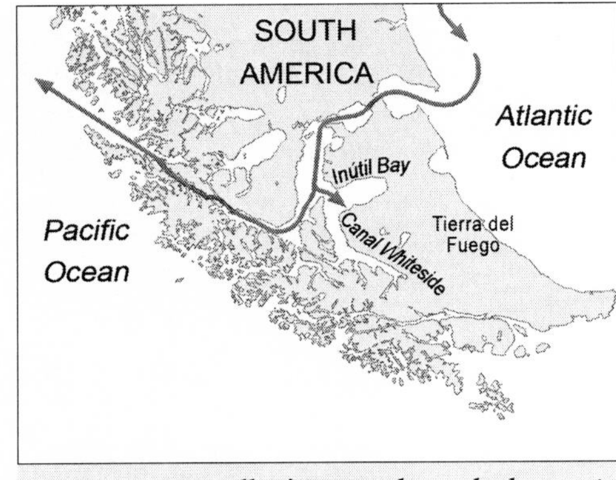

Figure 16. *Magellan's route through the strait. Note the fork in the waterway midway through. Magellan sent two of his ships in to explore the waterway now known as Inútil Bay and Canal Whiteside. Only one of the two ships, the Conception, returned and reported that this was a closed waterway lacking passage to the Pacific.*

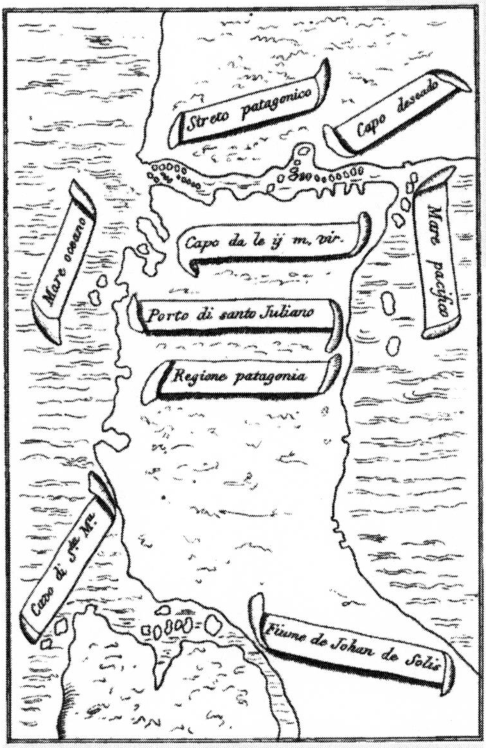

Figure 17. *Pigafetta's Map of Magellan's Strait. The map is drawn with a southern orientation, which accounts for the Pacific Ocean appearing to the right of the strait.*

This closed waterway extending off the strait was little more than a disappointment at the time of its discovery and may not have proved important from an historical perspective, but it was nevertheless included in Pigafetta's own charting of the strait. He drew it on a map included with his personal log of the epic journey (Fig. 17).

This bay would immediately influence maps and

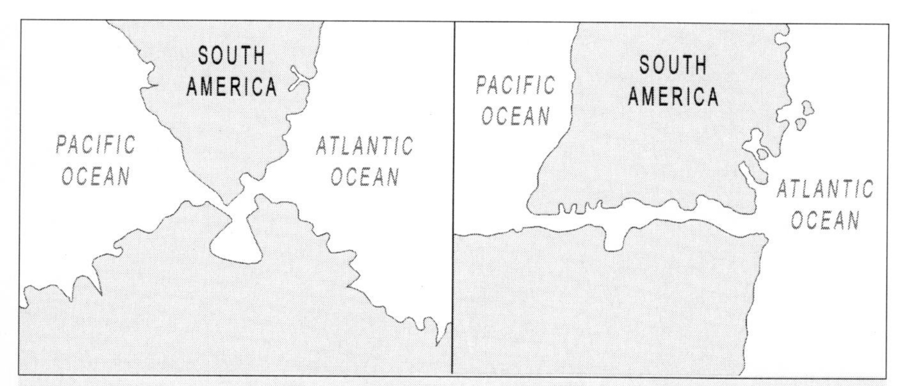

Figure 18. *The Strait of Magellan as portrayed on Finaeus' 1531 world map (left) alongside Antonio Pigafetta's map of the strait rotated to a northern orientation.*

globes of the period. Finaeus' map was one of those that included the deep bay penetrating into the southern shore of Magellan's strait. Interestingly, although described in his log as extending in a southeasterly direction, Pigafetta chose to depict it as a very basic U-shaped bay notched and aligned vertically into the strait's southern shore. Comparing this with Finaeus' 1531 depiction, it is clear that Finaeus' design was influenced directly by Pigafetta's visual depiction rather than the written description, as he drew a comparably shaped bay also aligned vertically (Fig. 18).

This point on Finaeus' design of the Antarctic continent corresponds to Atka Bay, which lies at the northernmost tip of Eastern Antarctica (Fig. 19). We can surmise that the cartographer responded to this unique detail of the strait and similar to Piri Re'is, who searched ancient maps to fill in missing gaps, did the same. The cartographer would no doubt have had a vast collection of maps both new and old at his disposal, so he set out searching for a landform that incorporated a similar bay.

In keeping with the possibility that Finaeus' map is an authentic map of Antarctica, it would seem that within this collection of maps the cartographer would have in his possession a genuine map of the continent. Further supporting this possibility, Finaeus' design strays from Pigafetta's overall design of a flat horizontal southern coastline for the strait by angling the bay's surrounding coastline downward, creating a coastline that more closely resembles Antarctica's Atka Bay and the

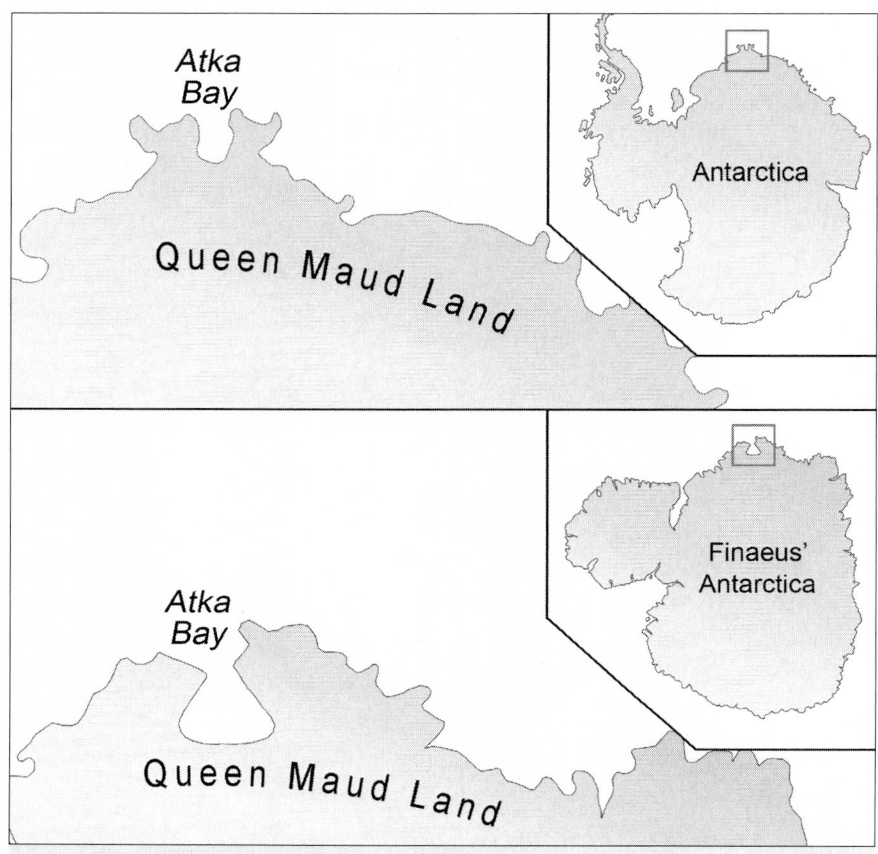

Figure 19. *Atka Bay, Antarctica (top), a U-shaped bay similar to Finaeus' depiction (below) in both look and location.*

surrounding Queen Maud Land. This suggests that the cartographer opted for a design dictated by a separate source map. If not, why not mimic either Pigafetta's description or depiction, instead of opting for a depiction that more accurately portrays the Antarctic northern coast?

While Atka Bay and Queen Maud Land appear to be the primary scaling point, we still need to determine the secondary point accounting for the continent's inclusion on a sixteenth-century globe as well as for its overscaling and misalignment. It is tempting to conclude that the source map identified and marked the South Pole and the continent was aligned accordingly. In fact, Hapgood concludes that the mapmaker attempted to align the continent to the South Pole, although it is not accurately positioned on the continent. Personally, I believed the crisscrossing of longitudinal delineations on an internal featureless landscape was more likely to be incidental and wanted to find a more convincing distinct geographic feature. It seemed more logical that the secondary scaling point would be a coastal feature such as a peninsula or even a bay like our primary point, Atka Bay, as these were objects more likely to be charted to a specific point on a map.

Yet in reviewing the three main maps at my disposal, I could not find a single shared coastal feature that set itself apart as a distinct secondary scaling point. There was one feature that did offer a bit of intrigue, a pair of small islands found on a world map produced in 1538 by Flemish cartographer Gerard Mercator and designated deserted islands on Finaeus' 1534 cordiform map—but Finaeus aligns them vertically while Mercator aligns them laterally.

Additionally, the islands seem to be locked onto the Tropic of Capricorn; the rendering of Western Antarctica hovers around 350 miles below the islands on Finaeus' 1534 map and Mercator's 1538 map distances Western Antarctica nearly twice that, some 600 miles south of the islands. There did not seem to be any correlation between the Antarctic continent and these islands.

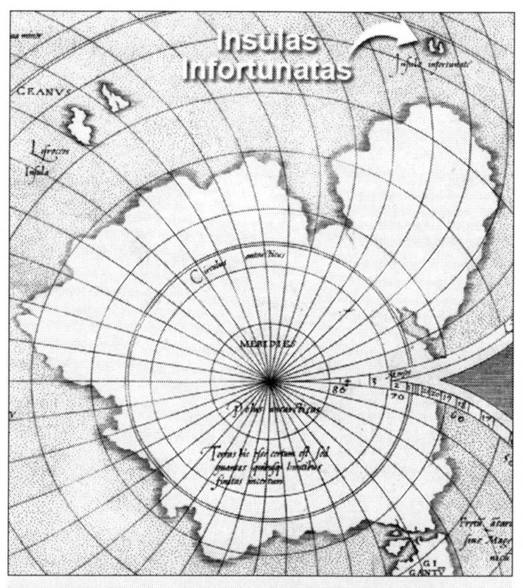

Figure 20. *Antoni Lafreri's 1564 replica of Mercator's World Map with two islands lying off the coast of its portrayal of Western Antarctica. An inscription below the islands identifies the islands as* "Insulas Infortunatas." *The Unfortunate Islands.*

Even more problematic, Finaeus' 1531 map, which was the oldest version of this Antarctic design among the set, did not have these islands charted at all. This suggested that the islands were discovered some time after the continent had already been introduced onto these early maps.

Only after exhausting every other possibility and being on the verge of giving up pursuing the map's authenticity did I decide to investigate these islands. I had very little expectation; I only wished to ensure that I concluded my research with no stone left unturned. The investigation began with an inscription placed just below Mercator's depiction of the mysterious island set that provides us with their name. The inscription is in Latin and reads, "*Insulas Infortunatas*": The Unfortunate Islands (Fig. 20).

A Tale of Two Islands

November 28, 1520 found Magellan crying tears of joy as he finally exited his strait into the Pacific Ocean. He had discovered the elusive western passage to the Orient. The historic undertaking was not without its sacrifices. Magellan had barely evaded a Portuguese naval detachment dispatched to seize him and his ships in order to prevent Spain from laying claim to an alternate route to the Spice Islands. He overcame mutiny aboard three of his five ships brought on by the rigors of the long uncertain voyage and the intense distrust of a Portuguese commander leading a Spanish expedition. He lost one ship, the Santiago, to grounding in a storm and he was soon to learn that another ship, the San Antonio, commanded by his nephew Alvarus Meschito, was also missing. Magellan had overcome the extreme trials involved in commanding an overly ambitious sixteenth-century expedition into the unknown and the discovery proved a more than fitting reward for his valiant effort. But the time for jubilation would prove short-lived. A harrowing journey still lay ahead of Magellan and crew as they exited the strait on the 28th of November and continued on the last leg of their journey across the Pacific with three remaining ships: the Trinidad, Victoria, and Concepción.

It was the simplest of plans: Sail northward several hundred miles, then proceed westward along a latitude close to that shared with the Moluccus Islands, now known as the Maluku Islands of Indonesia. There they would load their ships with valuable spices and supplies, then return home and report their grand discovery. There was only one problem with this plan. The route from the newfound strait to the Moluccus Islands was new and unfamiliar, and Magellan had grossly underestimated its distance. Magellan had imagined that the Moluccus Islands lay just on the other side of the New World rather than across a vast and unfamiliar Pacific Ocean. The consequences of this assumption would prove disastrous.

Magellan was so certain of the short distance remaining that he chose not to go ashore and replenish supplies before setting out across the Pacific. Nearly two grueling months after exiting the strait, the crew had not sighted any sign of land, and food supplies were growing dangerously low. On January 24, 1521, they finally discovered land in the form of a small island, but any excitement over the find quickly faded as the island proved to be barren—devoid of any life but trees and birds. They took soundings, but finding no bottom on which to set anchor, they continued on their voyage. They named this island San Pablo since it was discovered on the day of Saint Paul's conversion.

Sailing northwest, they traveled another 600 to 800 punishing miles before sighting more land. However, this again proved to be a desolate island with no place to drop anchor. They did manage to catch many sharks there and therefore named the island Isle of Tiburones and again continued on their way.

The extent of the crew's suffering was horrific. Outside of a few days dining on shark meat, the overextended voyage found them reduced to eating sawdust and stale biscuit crumbs soiled with rat feces and urine. They stripped ox hide strapping from the rigging and softened it in seawater for several days and ate that as well. Meanwhile, due to a lack of citrus, most of the crew suffered from scurvy. Their gums swelled to such a painful extent that they were unable to eat the little food they had. This eventually led to the death of 19 crewmembers. Dietary deficiencies also led to 25-30 cases of rickets, which softens the bones and causes the arms and legs to become severely disfigured. According to Pigafetta, it was not until March 16, 1521—3 months and 16 days after they had entered the Pacific—that they found relief. They arrived in the Philippines where they were able to replenish their supplies of wood, food, and fresh water.

The most unfortunate thing about Magellan's miscalculation of the size of the Pacific is that it caused him to set a course that swept right over and past the hundreds of habitable islands composing Polynesia. Had he known the true size of the Pacific, he certainly would have gone ashore and stockpiled supplies before setting out across it—but more importantly, he would have known there was no need to navigate northward so quickly to reach a latitude near that of the Moluccus Islands. A slower, more westerly ascent from the onset would have led him directly through Polynesia, where he might have laid claim to a bevy of new habitable islands teeming with food, water, and other supplies. Instead, Magellan chose a course that skimmed just above it, and in the process, discovered only two solitary uninhabitable islands in the vast stretch of the Pacific Ocean lying between the Strait of Magellan and the Philippines. The two barren islands that had nothing to offer the intrepid sailors but bitter disappointment were unceremoniously christened the Unfortunate Islands.

The Magellan Effect

Although it is a tragic tale, I could not help but be excited upon learning that these islands were discovered during Magellan's voyage. On this historic voyage, he added not one but two new geographic features to world maps: a strait with a southern bay and a set of islands in the middle of the Pacific. It was too perfect. Magellan had provided cartographers with two separate and distinct discoveries that would be two potential scaling points.

European cartographers, having received news of the successful voyage and discoveries, were eager to see the new lands portrayed on their latest maps and globes. It is true that Antarctica lay far south of these discoveries and indeed had yet to be discovered by European explorers, yet if an ancient map of Antarctica were somehow available and it had included Atka Bay as well as a set of two islands lying off the coast of Western Antarctica, it would have been difficult for these men to resist the possibility that these features represented the very same two features described by Magellan. Similar temptation overcame Piri Re'is when he adopted and scaled Ptolemy's landlocked ocean to his map.

Therefore, it seems more than reasonable that a cartographer, upon recognizing the similarities in the two features, would simply place Atka Bay at the tip of South America as a representation of Inútil Bay and stretch the continent across the map or globe until the two islands aligned with the location of the Unfortunate Islands.

Although to this day I still retain some doubts about the actual existence of an ancient Antarctic source map, at that moment I was so taken by the two islands' link to Magellan's voyage and the overall design of the continent that I was strongly convinced modern maps would confirm their existence off the coast of Western Antarctica. It was a eureka moment (albeit one that still required verification).

While I felt certain I would find that the islands existed, map after map resulted in disappointment. Online maps failed to display any sign of the islands. Fortunately, I had in my possession one large up-to-date world atlas. My excitement level was still high as I flipped the pages aside, finally revealing a large detailed spread of the Antarctic continent. My attention was instantly drawn to two well-defined islands. Nestled along the coast of Western Antarctica and enveloped within the Getz Ice Shelf under layers of ice and snow, there sat two islands: Siple and Carney (Fig. 21).

It was a truly astonishing find, but I was not out of the woods yet. If an ancient map was used as a template then these islands should appear on the first world map that incorporated the design. However, Finaeus' 1531 world map lacked these islands. His later 1534 map included the islands off the coast of Western Antarctica but oriented them vertically. Mercator's 1538 world map oriented the islands correctly but placed them too far off the coast of Western Antarctica. My only hope was that there was some earlier map that included the islands. This was an awkward predicament: I felt strongly that I was on the right track but did not feel comfortable hanging my hypothesis on the possible existence of a map predating Finaeus' 1531 version. I would also have to pin my proposal to the existence of a separate cartographer having introduced and enlarged the ancient design as Finaeus would likely not have omitted the islands from his 1531 world map if they played such a pivotal role in his selecting and scaling the ancient design to his map.

I searched for more sixteenth-century maps and globes exhibiting

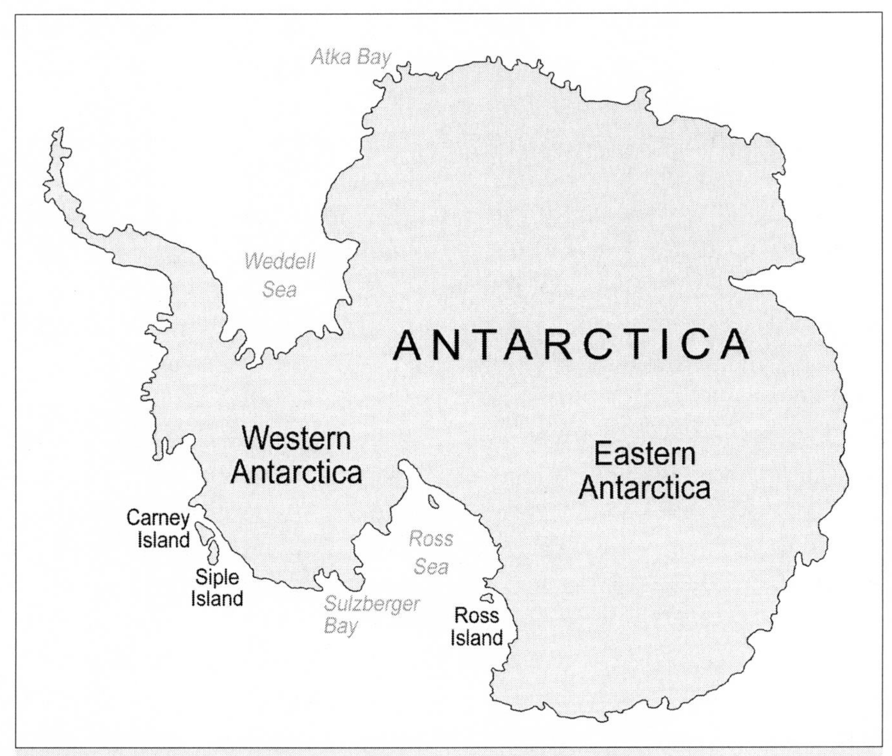

Figure 21. *Modern map of Antarctica, which reveals that islands corresponding to the Unfortunate Islands, Carney and Siple, actually do exist off the flattened coast of Western Antarctica.*

the same design; while I found quite a few, none predated Finaeus' 1531 map. Many maps of the sixteenth century are no longer with us, so it was highly feasible that the map I was looking for no longer existed.

As it turned out, I had run across the map I was looking for many times but had not realized it. While leafing through a copy of Hapgood's *Maps of the Ancient Sea Kings,* my attention was drawn to what I had thought were a series of varied projections of Finaeus' southern continent. I noticed that one of the maps differed slightly.

Hapgood had inserted the map into the set with seemingly little regard for its significance. The only reference to it in the book that I am aware of is in its caption, where he identifies the map as a portrayal of the Antarctic continent from Johannes Schöner's 1524 globe (Fig.

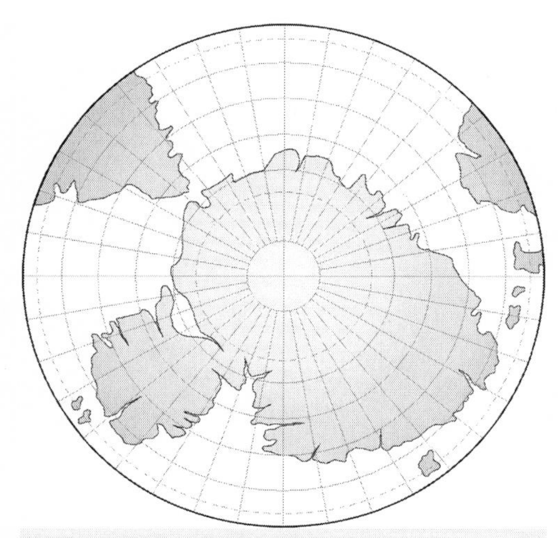

Figure 22. *Hapgood's facsimile of Schöner's 1524 globe. Unlike maps and globes that followed, this initial introduction of this Antarctic design includes the most accurate portrayal of Carney and Siple Islands as the author had predicted.*

22). But this map is incredibly significant. Not only was this the earliest depiction of the oversized Antarctic continent, it was created right on the heels of Magellan's voyage. Like the other depictions, the continent forms the southern shore of Magellan's strait with a bay similar to Atka Bay located in its midst. More importantly, the map not only included a depiction of the Unfortunate Islands, but they more accurately mimicked Antarctica's Siple and Carney Islands than later maps—just as I had predicted.

There is no mistaking the marked similarities between Schöner's depiction of the Unfortunate Islands and Antarctica's Siple and Carney Islands. And just as Schöner opts for a bay within the Strait of Magellan that more closely resembles Atka Bay than the waterway described by Pigafetta, he also ignores Pigafetta's description of the Unfortunate Islands and adopts a design that more closely resembles Siple and Carney Island in Antarctica. These discrepancies support the possibility that Schöner preferred a source that strayed from but approximated Pigafetta's description of the features.

According to Pigafetta's account and other varied accounts of the voyage, the Unfortunate Islands were located some 600 to 800 miles apart east to west. All accounts agree that there was also a separation of 4 to 6 degrees latitude; Pigafetta declares the island of San Pablo to be located at 15 degrees south latitude and the island of Tiburones to be 600 miles west of San Pablo at 9 degrees south latitude. Other sixteenth-century maps like the Hadji Ahmed Map of 1559 (Fig. 23) conform precisely to these parameters, while Schöner's map places the two

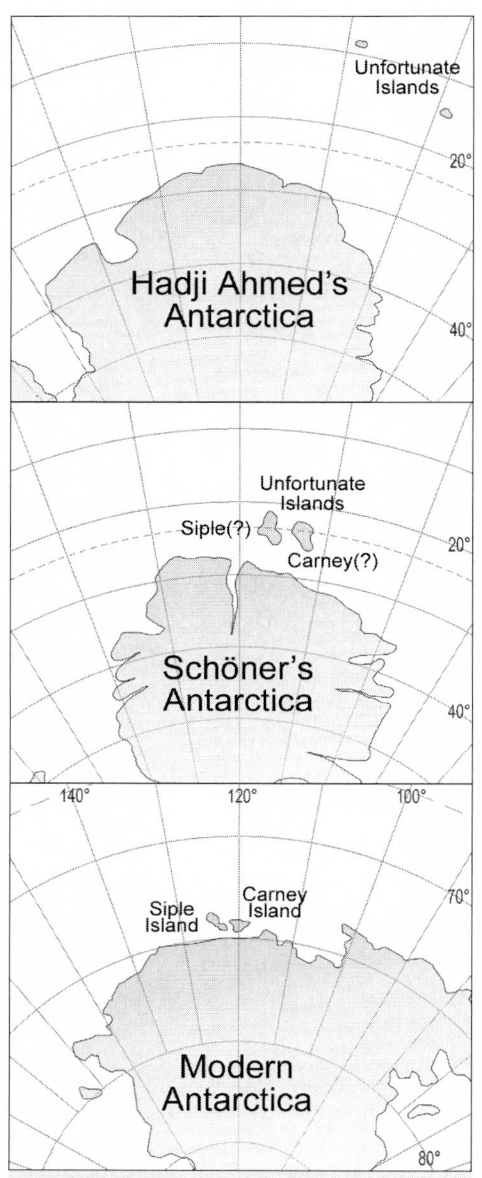

Figure 23. *Schöner (middle) portrays the islands more like Carney and Siple islands (bottom) than the actual Unfortunate Islands as described by Pigafetta (Hadji Ahmed Map top).*

islands on the same parallel and also disregards the stated 600-mile distance between islands, locating them a short 100 miles apart. The end result finds Schöner providing less a portrayal of the Unfortunate Islands and more a respectable portrayal of Siple and Carney, a tightly paired set of islands with a channel running between that is similarly proportioned and aligned. Schöner even accurately replicates the two islands' correlation to the Antarctic continent, placing the islands along and parallel to Western Antarctica's flattened westernmost coastline while offsetting the pair toward the northern end of the flat where the coastline begins to taper away in the direction of the Weddell Sea.

The evidence suggests that Schöner was driven by an ancient chart of Antarctica to place the whole of Western Antarctica far up into the Pacific in order to fit Siple and Carney Island to the approximate location of the Unfortunate Islands. Simultaneously, he scaled the remainder of the map so that Atka bay could also fit below the tip of South

America. This sheds light on why Schöner, no doubt aware of the incredible hardships Magellan and his crew suffered on their journey into an immense and vacuous ocean broken only by two inhospitable islands, would suddenly feel compelled to chart a large landmass sitting a tantalizing one hundred miles south of Magellan's route. It would almost seem a cruel gesture of irony on his part unless he sincerely believed the islands were associated with a nearby landmass. An ancient map of Antarctica with a lone pair of islands lying just off one of its shores would certainly have proved compelling.

Aside from the omission of the Palmer Peninsula and the apparent lack of an ice sheet, between Schöner, Finaeus, and Mercator we find not only accurate portrayals but also accurate positioning and alignment of: Atka Bay, the Weddell Sea, Coats Land, the Ross Sea, Ross Island, Sulzberger Bay, the Executive Committee Mountain Range, the Ellsworth Mountains, the Queen Maud Mountains, and now Carney and Siple Island.

Of course, we could still credit all this to extreme coincidence—until we ask ourselves how Schöner happened upon this design. We could argue that he either drew the continent from an existing design as discussed, or that it was simply the product of his imagination. Which of the two options would be the more responsible approach for a skilled sixteenth-century cartographer?

As we witnessed with the Piri Re'is map, the incorporation of older map designs was a process practiced by at least some cartographers during that same period. But to ascertain whether Schöner also used this process, it would require more proof. I thought it worthwhile to go directly to the source and look into Schöner's other works. It would be a decision that would lead to some completely unforeseen results.

CHAPTER 5

THE MAP AT THE BOTTOM OF THE WORLD

The sixteenth century was a period of heightened world exploration spurred on by the thriving spice trade and a major discovery by Christopher Columbus in the latter part of the previous century. In 1492, Columbus had sailed west from Europe in search of a shorter route to the spice-rich East Indies. Unfortunately, he fell short of this goal when he encountered a rather large obstacle that would eventually come to be known as the Americas. Regardless, the voyage and its surprising find captured the attention of Europeans intrigued by the adventurous tales of discovery. More expeditions in search of the elusive western sea passage followed and with the return of each expedition came word of new geographic discoveries, feeding the fire of exploration.

This fevered pace of exploration and discovery also drove a strong demand for charts and globes with up-to-date portrayals of the world. It is against this backdrop that we find cartographers like Johannes Schöner producing artistically rendered model globes depicting the latest geographic finds—and it is here that we begin our determination of his methodology in creating his Antarctic continent.

Schöner's Methodology for Cartographic Incorporation of New Discoveries

As we have seen, after Magellan's voyage Johannes Schöner incorporated the famous strait into his globe of 1524. However, since information on the strait's southern shore was lacking, it became the large Antarctic continent that graced this and many other maps that followed.

This was not the first time Schöner had had a go at depicting a southern continent. Schöner took his first stab at depicting the continent

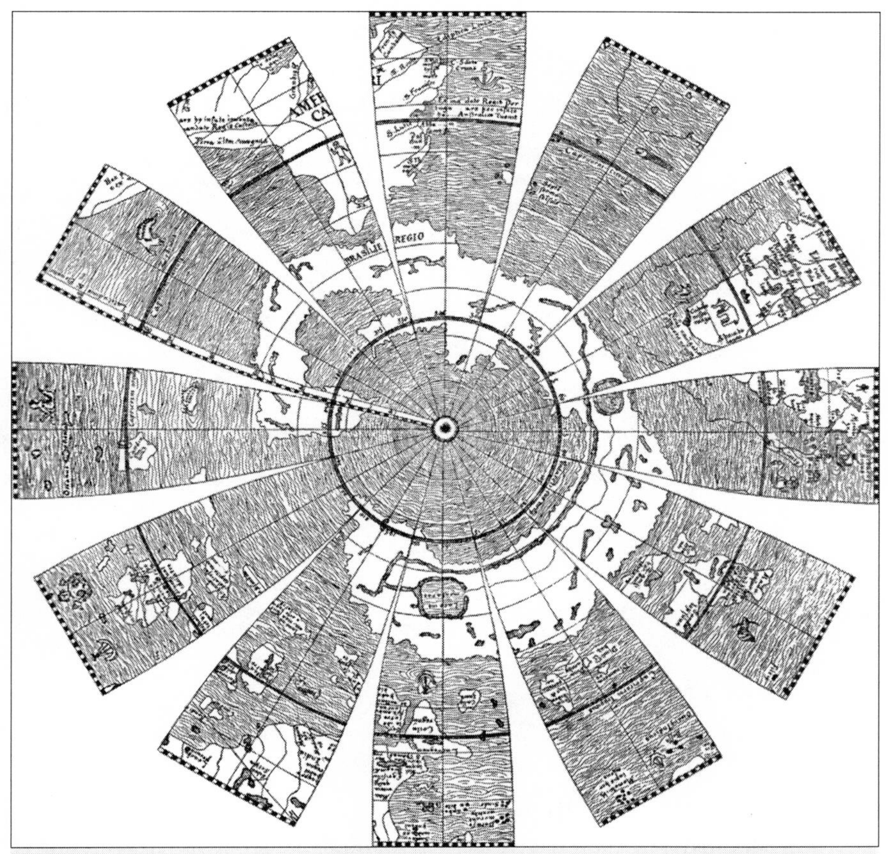

Figure 24. *Johannes Schöner's 1515 globe gores of the southern hemisphere, showing a peculiar version of a southern continent that forms a ring around the southern pole.*

on a globe he fashioned in 1515. The landmass as depicted on this globe (Fig. 24) has very little in common with its 1524 counterpart. About the only similarity we can see is that it is an oversized continent located just beyond the tip of South America, creating a narrow passage similar to the Strait of Magellan.

The inclusion of this strait has led some to suppose mistakenly that the strait had been discovered prior to Magellan's 1520 voyage, but this misconception is easily exposed under closer examination. The whole of Schöner's 1515 strait lies between 38 and 47 degrees latitude south, whereas the Strait of Magellan lies below the 52nd parallel. This locates Schöner's strait more than 350 miles north of the actual Strait of Magellan.

The reason for this misguided placement appears linked to both a pervasively optimistic belief that such a passage would be found and a misleading report that appeared in a German tract printed circa 1508 in Augsburg, *Copia der Newen Zeitung auss Presillg Landt* (New Tidings out of the Land of Brazil):

> *Learn also that on the twelfth day of the month of October, a ship from Brazil has come here, owing to its being short of provisions. The vessel had been equipped by Nono and Christopher de Haro, in partnership with others.*
>
> *Two of those ships were intended to explore and describe the country of Brazil, with the permission of the King of Portugal. In fact, they have given a description of an extent of coasts, from six to seven hundred leagues [1800 to 2100 miles], concerning which nothing was known before.*
>
> *They reached the Cape of Good Hope, which is a point extending into the ocean, very similar to, Nort Assril, and one degree still further. When they had attained the altitude of the fortieth degree, they found Brazil had a point extending into the sea. They have sailed around that point, and ascertained that the country lay, as in the south of Europe, entirely from east to west. It is as if one crossed the Strait of Gibraltar to go east in ranging the coast of Barbary.*
>
> *After they had navigated for nearly sixty leagues [180 miles] to*

round the Cape, they again sighted the continent on the other side, and steered towards the northwest. But a storm prevented them from making any headway. Driven away by the Tramontane, or north wind, they retraced their course, and returned to the country of Brazil.

The tract relates the account of a Portuguese sponsored expedition that explored over 1,800 miles of previously unexplored Brazilian coastline—Brazil being the name then applied to the whole of South America. We can deduce from the numbers provided that the exploration of new coastline began some 1,620 to 1,920 miles north of the 40th parallel, somewhere in the range of Sao Paulo and Rio de Janeiro, Brazil. The continuation of the sailors' route is plotted out on both Schöner's 1515 interpretation of the South American continent and a modern map of South America (Fig. 25). Schöner's sixteenth-century rendering of the South American coastline approximates the modern rendering as well as could be expected from a map of the time, but as soon as we pass beyond the 40th parallel we begin to see where his interpretation strays radically.

Schöner relies wholly on the account provided by the sailors who, undoubtedly elated over the possibility of having discovered the prized passage to the Pacific, related an overly optimistic description of the

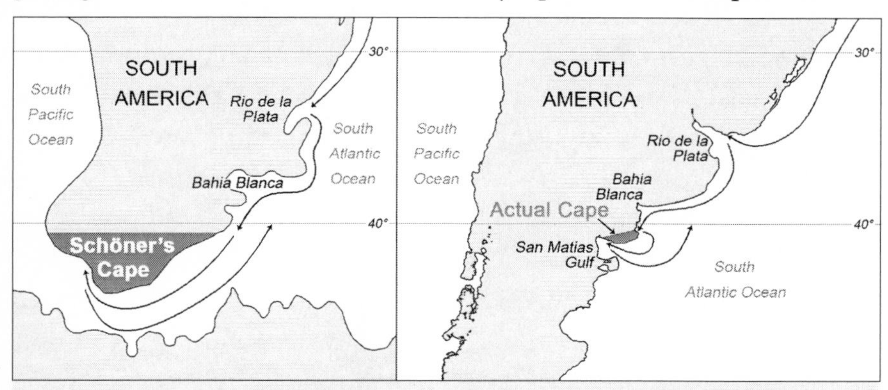

Figure 25. *The route taken by the Portuguese sailors laid out on Schöner's 1515 version of South America (left) reflecting Schöner's interpretation of the cape that extended into the sea beyond the 40th parallel. And the same route laid out on a modern map of the continent (right,) which clearly shows that the point extending beyond the 40th parallel was merely the northern shore of the San Matias Gulf.*

region. They had actually discovered the San Matias Gulf; this complies with the sailors' account of rounding a point of land 180 miles beyond the 40th parallel. That point appears to be the San Matias Gulf's convex northern shore. If we trace a course along the coastline from the 40th parallel, the shoreline begins to rise at around the 160-mile mark. At mile 180, the sailors noticed this continuing rise and believed they were heading up the western coast of the continent. Had they been able to sail another 40 miles, they would have spotted the closed western end of the bay, but they were deterred by a northwest wind that blew them down and out of the gulf. This wind would have directed them toward the gulf's southern shore and past the Valdez Peninsula; after spying this southern coast, they began to piece together the extent of their find.

While all they had truly witnessed was a waterway of undetermined length flanked by shorelines to the north and south, it did not prevent the sailors from embellishing with some of their own suppositions. And though they never *directly* state that the inlet was a through passage, they do strongly suggest it by equating their brief encounter with the bay as mirroring passage through the Strait of Gibraltar into the Mediterranean Sea. To whet the reader's imagination, they add that this new inlet was similar to traveling eastward through the Strait of Gibraltar "to go east in ranging the coast of Barbary," the vast North African coast. This is a strong implication that they had spotted an extensive coastline composing the strait's southern shore—although what they had actually seen was only a few miles of coastline.

Equipped with the misleading account, Schöner was ready to begin the process of incorporating the new discovery onto his globe. Based on the sailors' tale, he envisioned a South American continent that tapered to a point just beyond the 40th parallel, hovering above the coastline of a land of undetermined size thereby forming a strait between the Atlantic and Pacific oceans. Incorporating this newfound strait itself onto his 1515 globe was the easy part; based on so little detail, how would Schöner determine how to depict the land south of the strait? This leads us to Step 1 in Schöner's methodology.

Step 1: Referencing Ancient Source Maps

Determining the first step in Schöner's mapmaking process is the most crucial, as it could validate both his 1515 and 1524 Antarctic continents as designs based on ancient source maps rather than randomly contrived designs. Yet in looking at his 1515 globe, his Antarctic design looks like an object lesson in creative design.

Schöner renders the Antarctic continent as a large irregular C-shaped landmass that certainly looks a bit out of place on his globe when compared to the other continents and their more realistic geographic forms. The only real-world landmass that even comes close in appearance is an atoll, but an atoll approaching this size does not exist. (The largest existing atoll is the great Chagos Bank south of the Maldives, and Schöner's continent dwarfs it a thousandfold.) Yet it is within the realm of feasibility that Schöner may have misinterpreted and scaled a map of an actual atoll onto his globe. This would at the very least support the theory that he was referencing earlier existing source maps for inspiration.

After examining many atolls throughout the globe, I was unable to find one that approximated Schöner's design. There is one feature on the design that makes for a difficult match; a set of two prominent peninsulas located just clockwise of the opening in his C-shaped continent and extending into the interior body of water. However, these two peninsulas become a bit more accommodating if we readjust our sights from two peninsulas extending into a lagoon to two much larger peninsulas extending into a very sizable sea.

It was only after many days of studying the landmass that I keyed in on the peninsulas and finally recognized what I was looking at. It was both unexpected and mystifying. Due undoubtedly in part to the odd rendering of the landform, Schöner had managed to do the unthinkable. The two peninsulas actually depict two distinct and highly recognizable geographic features in the Mediterranean Sea: the lone set of prominent peninsulas, Italy and Greece. As we will shortly establish in full, Schöner had unwittingly affixed an entire ancient world map onto the bottom of his world globe. This will prove a fortunate mistake

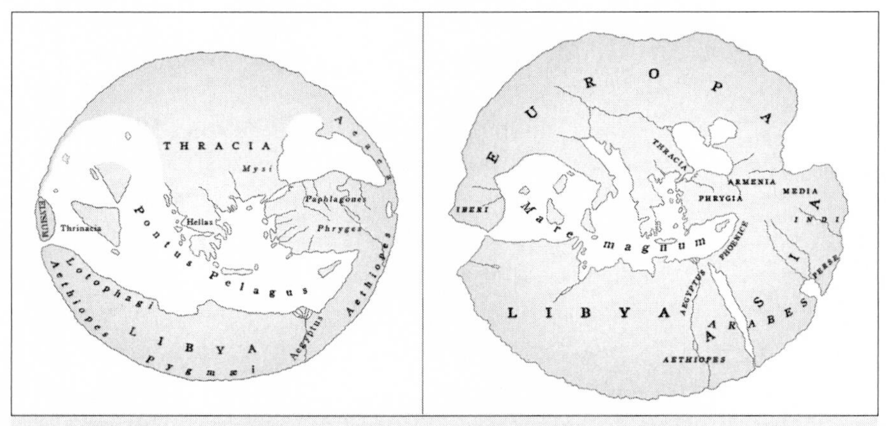

Figure 26. *Reconstructions of Homer's (left) and Hecataeus' (right) world maps.*

for us, as we now have solid proof that Schöner had indeed referenced and copied an ancient source map in designing his southern continent.

The overall design of this particular world map shares its pedigree with maps from Ancient Greece. Ancient Greek maps such as those of Hecataeus and Homer (Fig. 26) depicted the world as a circular disk with Europe, Asia, and Africa united in a singular circular band of land wrapped around a large inner sea, the Mediterranean. Beyond that, the whole of the world is surrounded by an outer ocean, with the Strait of Gibraltar located at the western end of the Mediterranean Sea forming the only passage between the inner sea and outer ocean. The combination of the Mediterranean Sea and Strait of Gibraltar effectively separates Europe from Africa to create a simple world map in the shape of a reverse "C."

To better visualize the similarities between the ancient Greek maps and Schöner's map, Figure 27 omits the extra detailing of inland bodies of water on the Hecataeus map. It breaks this ancient map down alongside Schöner's map into their three basic components: 1) the upper arm, Europe; 2) the lower arm, Africa; and 3) the two peninsulas, Italy and Greece.

The point where the southern coast of Turkey veers perpendicularly into the Mediterranean effectively splits each map into two similarly sized halves—Europe composing the northern half and Africa the southern half.

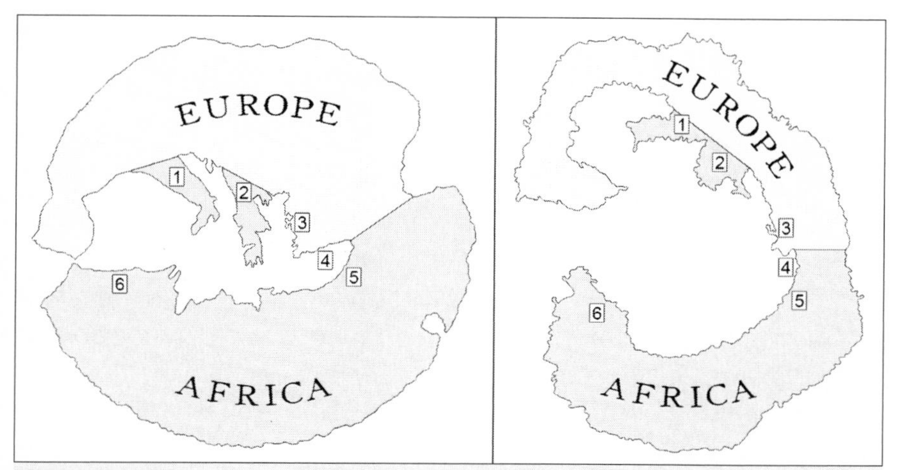

Figure 27. *Breakdown of Hecataeus (left) and Schöner's (right) world maps into 3 basic components: Europe, Africa, and the dual peninsulas of 1) Italy and 2 Greece, extending from Europe into the Mediterranean Sea. Other shared elements include: 3) a small peninsula and gulf along the western coast of Turkey representing the ancient port city of Smyrna; 4) the southern coast of Turkey cantilevered out over Africa; 5) the curved transition from Egypt to Syria; 6) the elevated coastline of Western Africa; and finally, the similar arrangement and relative proportions of all these elements within their respective portrayals.*

In evaluating the lower half, both maps round out the naturally squared corner between Israel and Egypt, creating a continuous coastline that sweeps downward beneath Turkey and across North Africa. Following this coastline westward, the last third exhibits a significant abrupt rise, a reasonable portrayal of the stepped appearance of the North African coast. On a modern map, the northern coasts of Egypt and Libya remain on a level that never rises beyond the 33rd parallel, but toward the west the coastline rises up into the Mediterranean at Tunisia. It levels off around the 36th parallel with a coastline comprising the northern coasts of Tunisia, Algeria, and Morocco. Schöner's rendering may be lacking in some details but allowing for the overall primitive nature of this world map, the similarities it shares with the North African coastline prove fairly accurate.

Turning our attention to the upper arm we see a similar arc, but Schöner's European arm does deviate significantly by angling the Iberian Peninsula inward. It does, however, depict a point (though exaggerated) at the very end of the country that likely represents a point of land extending off the Iberian peninsula's far end just beyond the town of Portimão, Portugal and out to the town of Sagres.

Moving clockwise we find the two imposing peninsulas portrayed accurately protruding from the northern arm of Schöner's 1515 landmass into the interior sea. They lie adjacent to each other with the thin rectangular peninsula representing Italy to the west. It includes two opposing peninsulas extending off its farthest extremity, fashioning a rather primitive version of the toe and heel of the signature "Boot of Italy" (Fig. 28).

The peninsula depicting Greece to the east is rendered in a rather primitive fashion as well yet is detailed sufficiently to define the location of Athens. Athens is located midway down a narrow strip of land extending eastward off the main trunk. Alongside the site of Athens is a small narrow nub of land jutting back toward the west. On Schöner's map, this small feature terminates abruptly where it should normally form the Corinthian Isthmus that connects to the large Peloponnesos Peninsula. This omission could be attributed to many factors, among them the poor condition of the source map or simply an oversight by an ancient cartographer or copyist.

The final recognizable detail pertains to Turkey. It is first worth noting that Turkey's southern coastline, although a bit condensed, exhibits an undulating shoreline very similar to its modern-day coastline. While this feature alone may not be overly convincing in validating this as a portrayal of Turkey, taken in tandem with the detailing of a small hook-shaped peninsula extending off the western coast, the combined features prove sufficient. The hook-shaped peninsula is a rather accurate rendition of the peninsula that bounds the Gulf of Izmir. Not only is the peninsula correctly located east of the representation of Greece, but it is also accurately aligned with its point paralleling the coast in a counterclockwise direction. The gulf itself accurately conforms to the actual gulf's 40-mile deep by 20-mile wide proportions and even appears to add the little bump of land that projects out from the lower end of the gulf. The fact that this gulf is one of the more accurately defined features of the map reflects the importance of the ancient city

Figure 28. *The renderings of Italy, Greece, and Turkey on Schöner's 1515 landmass down the left side with their real-world counterparts to the right. All three of Schöner's peninsulas not only share key characteristics with their counterparts, but more significantly, they are laid out in correct sequence on his map. The evidence dictates that Schöner's landmass was copied from an ancient world map.*

of Smyrna, a settlement inside the gulf that existed as a prominent city port under both Greek and Roman rule.

Presented with the evidence thus far, it is reasonable to assume that indeed an ancient world map occupies the larger portion of Schöner's 1515 southern hemisphere. In my experience, every reverse C-shaped landform exhibiting two peninsulas extending from their upper arm into the central void has been an ancient world map, so I was not about to let Schöner's errant placement of this comparatively designed landform on the bottom of his globe allow me to write the similarities off as coincidence. Especially in light of the fact that Piri Re'is similarly scaled features from ancient maps, it still seemed far more practical that Schöner followed this practice as opposed to freestyling from scratch.

If this is indeed a misplaced ancient world map, who do we credit as the original source? The Greeks? No—the overall design of the map negates this possibility. The Greeks maintained a Greco-centric view of the world that they projected onto their simple circular world maps. This customary practice entailed locating Greece in the geographic center of the landform. (Reference the Hecataeus and Homer maps, Fig. 26.) Schöner's map clearly does not conform to this rule of design, which would rule out Greek origins.

We find the key to the source map's origins in a feature stretching across the width of Schöner's Africa in the form of the map's lone inland waterway. This unusual water feature appears as a long, thin, undulating channel terminated at both ends by large mountain lakes while to the north a lengthy mountain range runs parallel to the length of the waterway. When I first realized that I was looking at an ancient world map, I saw this waterway and the mountain range above it as a major hurdle since no such waterway and mountain range existed in Africa. Much to my surprise, these features would prove to be the evidence I needed to verify that I had indeed discovered an ancient world map—and more specifically, an ancient Roman map. Although there are no such water features even remotely similar in Africa, this design follows ancient Rome's concept of one of the most historically renowned rivers in the world: the Nile.

Figure 29 displays a modern reconstruction of one of ancient Rome's most famous maps, Agrippa's Orbis Terrarum, which was a large display map completed around 20 CE. Copies of the map were distributed throughout the ancient empire and continued to appear in medieval Europe, whose mapmakers referenced them to design the *mappae mundi*, medieval maps of the world. Copies of this Roman map eventually disappeared; therefore, modern reconstructions like the one in Figure 29 are only approximations of the map and base their design upon a combination of geographic information gleaned from ancient historians as well as the *mappae mundi* derived from the Roman original.

One of the more noticeable aspects of this reconstruction is its

Figure 29. *A modern reconstruction of Agrippa's Orbis Terrarum, which like Schöner's map depicts a lateral landlocked waterway in Africa that is terminated at each end by large lakes.*

orientation, which borrows from the medieval practice of aligning east toward the top of the map, a cartographic practice for which the word "orient" acquired the alternate definition "to align." The feature we want to focus on, however, is the small laterally arcing waterway positioned in the middle of Africa. It is not nearly as imposing as Schöner's lengthy waterway and exhibits a tighter, smoother undulation, but like Schöner's version, it sets itself apart not only as the only waterway that is completely landlocked but also the only waterway terminating in large lakes at both ends. All other depicted rivers empty into seas or the outer ocean. Like Schöner's map, this unusual waterway effectively divides the continent of Africa into two regions, north and south. If there were any doubts regarding Schöner's southern landmass having been based on an ancient world map, this feature should fully dispel those doubts.

This water feature actually maintains ancient Rome's basic misconception that the source of the Nile originated in a West African mountain lake and flowed eastward across the continent. Pliny the Elder, a first-century Roman historian, writes the following with regard to the Nile in his work *Natural History*:

> *It rises ... in a mountain of Lower Mauritania, not far from the ocean; immediately after which it forms a lake of standing water, which bears the name of Nilides ... Pouring forth from this lake, the river disdains to flow through arid and sandy deserts, and for a distance of several days' journey conceals itself; after which it bursts forth at another lake of greater magnitude in the country of the Massaesyli, a people of Mauritania Caesariensis.*
>
> *It then buries itself once again in the sands of the desert, and remains concealed for a distance of twenty days' journey, till it has reached the confines of Aethiopia. Here ... it again emerges ... forming the boundary-line between Africa and Aethiopia.*[3]

Pliny's account varies a bit from the river's layout on the reconstruction but essentially shares the concept of a source lake in Mauritania, western Africa, which feeds an eastward-flowing Nile River. In line with Pliny's description, this waterway extends across the

[3] Pliny, *Natural History* 5.10

continent to form a boundary between Africa to the north and Ethiopia to the south. In eastern Africa, this misconceived Upper Nile ceases its aboveground progress and supposedly empties into an underground river, where it continues to flow until it rises one last time as the Nile River with which we are familiar. From there it flows unabated above ground until eventually emptying into the Mediterranean Sea.

The modern reconstruction opts for a conservative portrayal of this landlocked waterway when compared to Schöner's depiction, but in referencing maps that are direct derivatives of Agrippa's map, it seems that Schöner's map may indeed be the more accurate depiction. One such map is the Hereford *Mappa Mundi* (Fig. 30).

The Hereford map, preserved in England's Hereford Cathedral, is a medieval map whose origin dates back to circa 1290 CE. Like Agrippa's map, it is a large map intended for public display, measuring approximately 62" tall by 52" wide including its decorative pentagonal

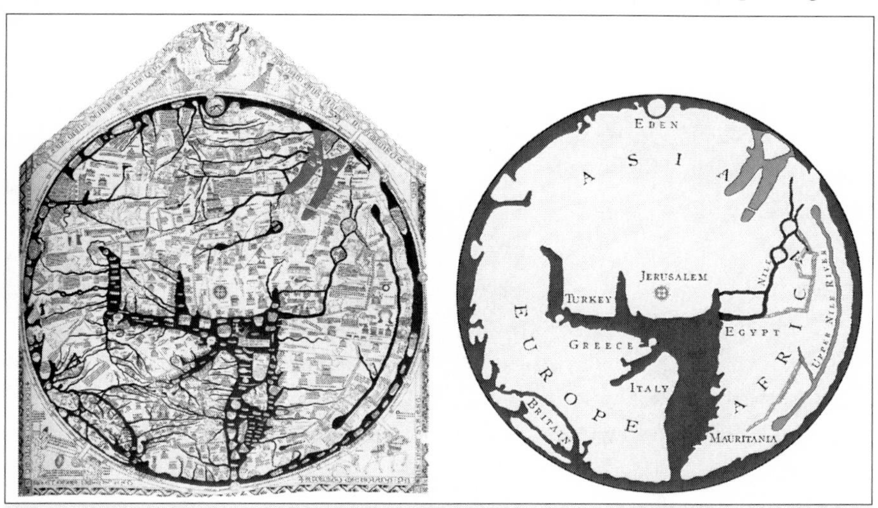

Figure 30. *The Hereford* Mappa Mundi *(left), perhaps the most renowned of the* mappae mundi, *alongside a stripped-down reproduction. Like the reconstruction of Agrippa's map, it incorporates the mysterious landlocked waterway spanning the width of Africa. Also of note is the lateral mountain range paralleling the waterway to the north. Signature features distinguishing the map from Greek maps are the city of Jerusalem positioned at the map's center and the Garden of Eden rendered as a circular island in the east.*

border. The map itself is a circular rendering, similar to ancient Greek design, but it employs the cartographic practice of orienting east toward its top. European cartography does add its own unique stamp on the circular map design with elements reflecting a medieval Europe that had transitioned into a Christian society. Taking its cue from Greek map design and its concept of cartographic centricity, the Hereford and other *mappae mundi* adopted a Christocentric design, locating the holy city of Jerusalem at the map's center. This radical design decision countered the practice of Greco-centricity with the preferred adherence to a literal translation of Ezekiel 5:5, "This is what the Sovereign LORD says: This is Jerusalem, which I have set in the center of the nations, with countries all around her."

Along with the requisite place names, the map is littered with inscriptions of varying lengths providing detailed information about the regions in which they are inscribed. While the majority of these inscriptions appear sourced from pagan authors exposing the map's Roman influence, at least twenty inscriptions are included on the map that further reflect Europe's Christian influence. This influence extends to the map's inclusion of a representation of the Garden of Eden near the top, as well as an image of Jesus being attended to by angels that adorns the upper portion of the map's border. The map also exaggerates the size of Palestine, allowing space for further Christian detailing such as an image of the walled city of Jerusalem with Christ's crucifixion drawn just above it.

The faded and discolored appearance of the map belies its original beauty. The original detailing was certainly very stunning with the surrounding ocean and seas colored green, red coloring applied to the Red Sea and Persian Gulf, and the numerous inland lakes and waterways scattered about the map differentiated with a deep blue. And most importantly, we find one of these blue waterways cutting a wide arc across the continent of Africa, which like Schöner's map, is terminated at both ends by large lakes. Also mirroring Schöner's map is a mountain range paralleling the waterway to the north, with the slight difference that Schöner's mountain range is rendered one continuous length extending well beyond the water feature in the east, while the Hereford mountain range is composed of two lengths of mountains, the eastern extremity terminating near the eastern end of the waterway.

So while the modern reconstruction in Figure 29 assumes the original Roman map portrayed a much smaller mythical Upper Nile, based on the Hereford Mappa Mundi it would seem likely that the waterway was a far more prominent feature similar to the portrayal in Schöner's map (Fig. 31). This is a very common portrayal of the feature shared by many of the mappae mundi—such as the tenth-century Cottonian, eleventh-century Isidorean, twelfth-century Henry of Mainz and Liber Floridus, and thirteenth-century Ebstorf maps, to name a few. Sometimes these maps portray the waterway as terminated by lakes at both ends and sometimes at only the western end, but all these maps, like Schöner's, show the unique waterway stretching across the width of Africa and terminating inland without ever emptying into the outer ocean or inner sea.

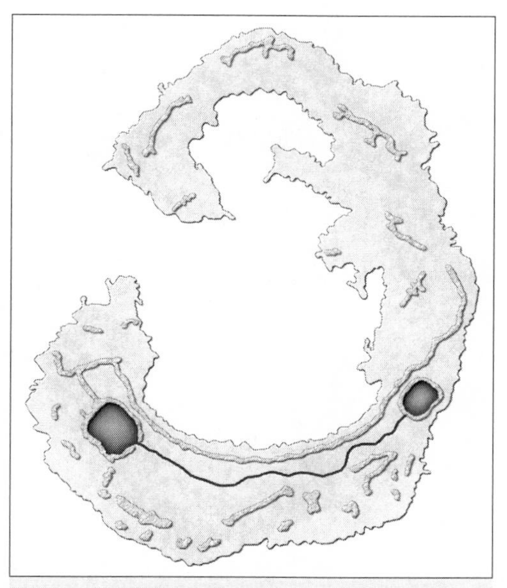

Figure 31. *Schöner's 1515 southern continent exhibiting a unique lateral landlocked waterway terminated by lakes on each end with a mountain range paralleling it to the north. It is a feature set only found here and on reverse C-shaped world maps of the Middle Ages. More importantly, this feature set is always found on the lower half of the 'C' opposite two prominent peninsulas.*

There were also some minor attributes that I initially overlooked on the design of Schöner's waterway that would have made it even clearer that Schöner was basing his design on a Roman concept of the Nile. As mentioned earlier, Pliny believed the western lake was the source of the Nile and that toward the east the waterway traveled underground. I had seen that the lakes at either end of Schöner's waterway were each surrounded by a ring of mountains; I had failed to realize that there were subtle but significant differences. The first difference was that Schöner

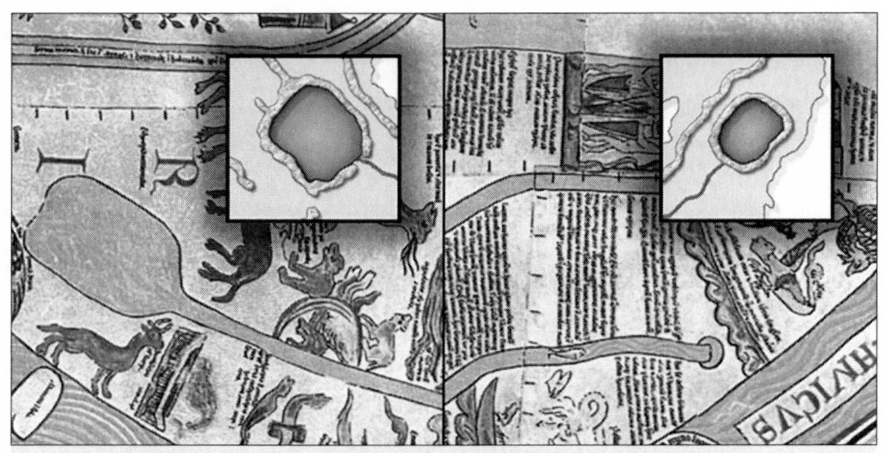

Figure 32. *Ebstorf western (left) and eastern lakes (right) with insets of corresponding lakes from Schöner's 1515 world globe. Not only do both maps depict the western lake as the larger of the two lakes that opens out into the lateral waterway, but Schöner completely encloses the eastern waterway with mountains, making it very clear that the waterway flowed underground beneath the mountains in the east. Similarly, the Ebstorf map truncates the waterway over the eastern lake to convey the waterway's sudden underground transition.*

drew the western lake as the larger of the two lakes. This is important as Pliny spoke of the western lake in Mauritania as being a "lake of greater magnitude." While the Hereford map portrays both lakes nearly equal in size, most *mappae mundi* depict the western lake as the larger of the two, the Ebstorf being one such map.

The other subtle detail that I had originally overlooked confirmed that Schöner was depicting the western lake as the source lake. In Figure 32, you can see that Schöner draws an opening through the mountains surrounding the western lake (left inset), but there is no such opening in the eastern lake's perimeter (right inset). It is clear that the lakes are drawn to portray the waterway emptying out through an opening in the western mountain lake, while in the east Schöner's waterway is drawn in such a way that the waterway must be directed beneath the mountains, or underground, to feed the enclosed lake. The Ebstorf map in the backdrop similarly conveys the concept of the western lake as free flowing, while the waterway is truncated over the smaller eastern lake.

This is another cartographer's effort to depict the waterway dropping underground in the east.

No doubt should remain that Schöner's southern landmass is a transplanted ancient Roman world map. Not only are landlocked waterways a rarity on ancient maps, but depictions of underground waterways are far rarer. To share a unique west-to-east flowing landlocked waterway terminated on both ends with lakes that are portrayed with this underground feature in the east defies coincidence.

The map's Roman heritage is further supported by the line of mountains Schöner portrays lying to the north of the waterway, which have no counterpart in real-life Africa. While the Hereford map shows this as a divided mountain range, ancient Roman maps portray it as a continuous chain spanning the entire width of the African continent, precisely as shown on Schöner's map.

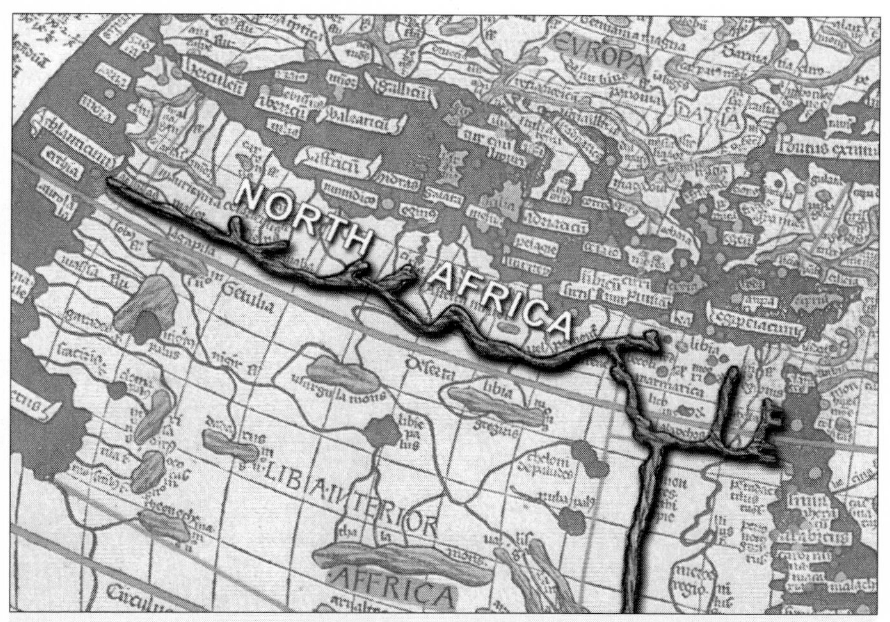

Figure 33. *North Africa as presented on Ptolemy's World Map, found within a 15th-century edition of Claudius Ptolemy's* The Geographia *(See Fig. 8). Note the length of mountains extending the full width of North Africa, leaving only a narrow strip of land to the north. Schöner's map similarly strings a mountain range just inside the coast of North Africa.*

Claudius Ptolemy's world map addressed earlier in this work (Fig. 8, p. 15) is one of those maps. Although Ptolemy employs a modified spherical projection that differs significantly from Schöner's less technical design, Ptolemy's map also includes a vast mountain range spanning the full width of the African continent (Fig. 33), sectioning off a very lean region in the north and leaving a much deeper one to the south.

This is a common Roman design concept; we can establish this fact with the second Roman map offered for our review, the *Tabula Peutingeriana* (Fig. 34). The *Tabula Peutingeriana* or Peutinger Table, is a replica of a first-century Roman *itinerarium*, essentially a road map of the Roman Empire believed to have been referenced in the making of Agrippa's *Orbis Terrarum*. It was discovered in 1494 in a library in Worms, Germany, and derives its name from Konrad Peutinger, the

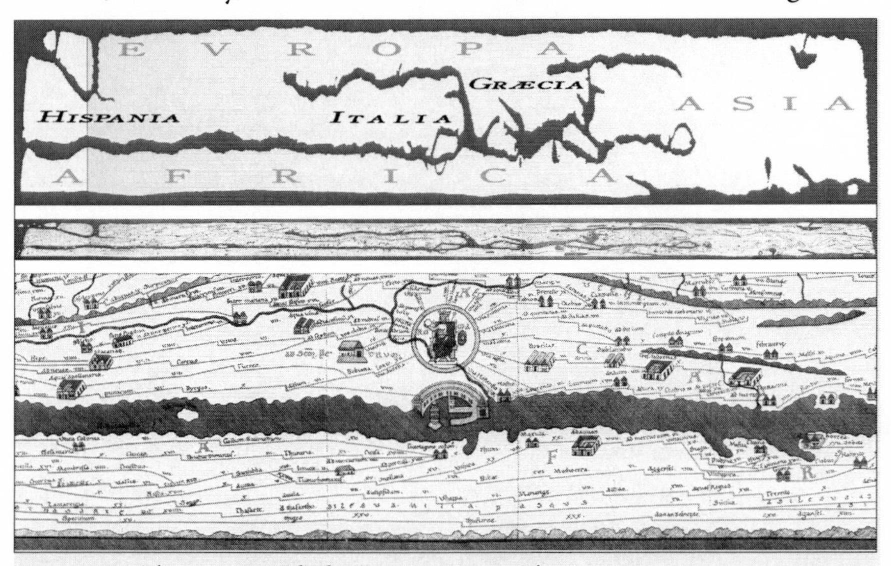

Figure 34. *The Roman* Tabula Peutingeriana. *The top image is a reconstruction stretched vertically for perspective while the middle is an actual scaled image of the map. The bottom image is an enlarged detailed section of the map. A long unbroken chain of mountains, which can be seen lining the map's bottom edge, marks the point where much of the African continent has been sheared away beneath. Like Ptolemy's and Schöner's maps, the mountains span the width of Africa, dividing a lean North Africa from the vast southern portion of the continent.*

man who eventually acquired the map in 1508. The Peutinger Table is another reverse C-shaped map but in rectangular form. It is composed of Europe, Asia, and like Ptolemy's and Schöner's maps, depicts northern Africa above a mountain range extending the full width of the continent. In fact, the mountain range lines the bottom of the map as the majority of the African continent has been cropped (apparently to facilitate the narrow form). The consistent depiction of a nonexistent continent-wide mountain range just inside the North African coast on all three maps further confirms that Schöner's map is based on a Roman world map.

Another significant aspect of the Peutinger Table is the map's deliberate distortion. In addition to shearing off most of the African continent, the table constricts and flattens both the Mediterranean Sea and the continents and also folds the Italian Peninsula in on its side so that it is pointing eastward toward a shortened Greek Peninsula. The purpose of these and other geographic distortions are clearly to keep the table within the confines of an easily transportable form, only 13½ inches in height and 22 feet of scrollable length. The distortion of the Mediterranean Sea would have been of little concern since the table was not intended for naval navigation. With its numerous roads drawn throughout, along with inscribed measurements revealing the distance of travel between key areas of the Roman Empire, the map was clearly intended for use by Romans traveling overland.

Like the Peutinger Table, Schöner's map deliberately distorts geographic features to achieve a particular design. Schöner replicates the Peutinger Table's placement of the Italian Peninsula on its side but directs it in the opposite direction away from the Greek Peninsula and greatly enlarges the Mediterranean. This distortion results in a large, imposing central circular version of the Mediterranean Sea delineated by Italy's altered eastern coast, Turkey's western coast, and the elevated northwest coast of Africa (Fig. 35).

While the Peutinger Table was distorted to facilitate ease of portability, Agrippa's was a large map used for public display, similar to the Hereford *Mappa Mundi.* So how would Schöner's design with its distortion of the Mediterranean find itself suited to this capacity? It is important to understand that the creation of the map was intended to record and display Agrippa's extensive and meticulous survey of the known world. As Pliny confirms, "Agrippa, a man of such extraordinary diligence, and one who bestowed so much care on his subject, when he proposed to place before the eyes of the world a survey of that world."[4]

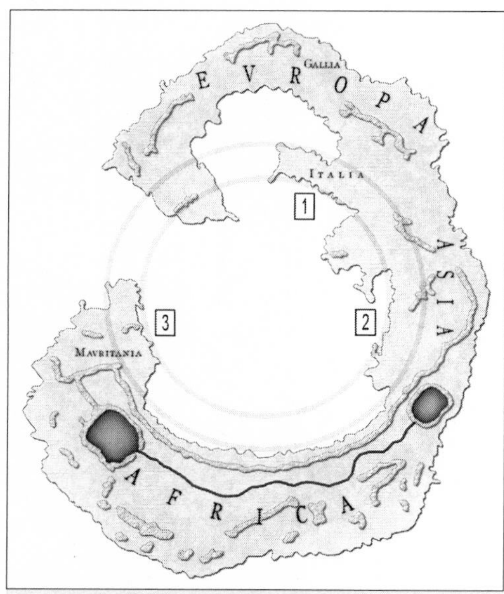

Figure 35. *Schöner's map with rings superimposed to demonstrate the concentric design of the Mediterranean Sea. Defining the innermost ring is 1) the eastern coast of Italy, 2) the western coast of Turkey, and 3) the elevated northwestern coast of Africa.*

This leaves us with a bit of a conundrum. Why would Agrippa feature the sea so prominently in the middle of his map when the map's true purpose was to survey the land? Assuming that Schöner's design is a true representation of Agrippa's world map, we should expect that the purposely oversized, rounded sea would incorporate content that augmented the map and Agrippa's land survey.

There is reason to believe that this central circular portion of the map was designed to house an extensive commentary detailing the known world depicted in the surrounding map. It is widely believed that Agrippa provided some form of commentary with his map that has since been lost. This is based in part on several passages in Pliny's *Natural History* that refer to apparent statements Agrippa made about his survey—about geographic information not conveyed through basic map elements and imagery alone. For example, regarding the inhabitants of one region Pliny writes, "M. Agrippa supposed that all this coast was peopled by colonists of Punic origin."[5] Regarding the inaccessibility of

[4] Pliny, *Natural History* 3.3

[5] Ibid., 3.3

certain geographic regions, "Agrippa states that the whole of this coast, inaccessible from rocks of an immense height, is four hundred and twenty-five miles in length, beginning from the river Casius."[6]

Additionally, the *mappae mundi* borrowed their pagan commentary from a Roman source, which was most likely Agrippa's map. This is not surprising considering the *mappae mundi* borrowed much of their design from his map as well.

The migration from a centrally located commentary suggested by Schöner's circular interior to the *mappae mundi's* inland distributive commentary is a very logical progression when placed in the context of the period. It would have been the direct result of Christocentricity, the desire by medieval cartographers to place the Christian holy city of Jerusalem at the map's center. Effecting this change required cartographers to expand Asia inward toward the map's center. If we consider Agrippa's design to have been similar to Schöner's in having a centrally located commentary, the inward extension of Asia would have displaced roughly half of the inner circle along with half the map's commentary. To retain the commentary and maintain Jerusalem at the map's center, the medieval cartographers had little option but to transfer the displaced commentary onto the enlarged Asian continent. In turn, maintaining consistency of design throughout the map would have further dictated the expansion of the European and African continents into the remaining half of the inner circle and similar redistribution of commentary over these enlarged continents.

Interestingly, the Psalter Mappa Mundi (Fig. 36), a late thirteenth-century map, demonstrates the feasibility and effectiveness of Schöner's design and the benefits of a consolidated commentary. The Psalter map measures a mere 3¼ inches in diameter, allowing it to fit inside the small book of Psalms from which it derives its name. It is similar to other mappae mundi in design but, because of its small size, is limited to including only a fraction of the place names and completely devoid of the commentary normally etched onto larger mappae mundi. As a workaround, a separate reduced commentary is located on the reverse side of the page, where it is inscribed within a circular frame, providing

---

[6] Pliny, *Natural History* 6.15

Figure 36. *The Psalter Map incorporated into a small book of Psalms with the map on the fore page (left) and commentary consolidated on the back (right), splitting the standard* mappa mundi *design into two logical and practical components.*

a rough glimpse at the way in which Agrippa's inner circle of text may have once appeared. The commentary's circular frame is actually a tripartite map, a medieval creation that is known as a T-O map because of the way it represents the world as a simple circle divided into three parts by a 'T'. The three resultant divisions represent the continents of Europe, Asia, and Africa. The Psalter map fills these three continental frames with commentary apportioned accordingly. Hence, on the fore map we find the now-familiar Upper Nile spanning the African continent with depictions of odd-looking imagined humanoids lining its southern shore, while within the African section of the tripartite map we find associated commentary referencing these "Ethiopian monsters."

The Psalter dual part map not only proves a logical and efficient means for preserving both the design of the *mappa mundi* and the commentary in small form; it also provides an intuitive device for referencing commentary without the need to scour the main map. Such a device becomes more practical when applied to a very large map. Assuming that Agrippa's map was much larger than the Hereford *Mappa Mundi* as it was originally intended for public display on the wall of a Roman portico, the large porch with column-supported cover located at the entrance of a building, it would have proved an extreme convenience to the viewer to access the map's extensive commentary consolidated within the confines of the map's circular rendering of the Mediterranean Sea.

At this point there is no doubt that we have indeed discovered an ancient Roman world map at the bottom of Schöner's 1515 world globe. All that remains for us to do is reconstruct the map to its original appearance. We will begin by attempting to establish the map's original orientation. The map's orientation has long been disputed among scholars, with north, south, and east all considered viable possibilities, but Schöner's map appears to confirm a northern orientation, much like the Roman Peutinger and Ptolemy maps. We can deduce this from the internal symmetrical geometric framework around which the map was built.

Artistic paintings often have the general detailing and layout sketched onto the canvas as guides before the paint is applied; so too was the case with ancient maps. While Ptolemaic maps benefited from a grid composed of latitude and longitude, these grids were not implemented on earlier Roman maps as evidenced by the Peutinger Table. Yet it appears that Agrippa's map does incorporate a structural guide. The map uses simple geometry to divide the Mediterranean Sea into three significant zones around which the three continents are wrapped.

Where is this geometric framework? It begins with a set of concentric circles. Figure 37 reveals four of these. The innermost circle (A) is formed by the coasts of eastern Italy, Turkey, and Northwest Africa. This represents the largest section of the Mediterranean Sea, Zone 2, and as discussed earlier this was most likely used to accommodate Agrippa's commentary. The next ring (B) runs along the opposite, western shore of the Italian Peninsula as well as along a short peninsula extending off Western Africa. (Note both peninsulas fit within the band created by circles A and B.) A third ring (C) is defined by the combined coasts of

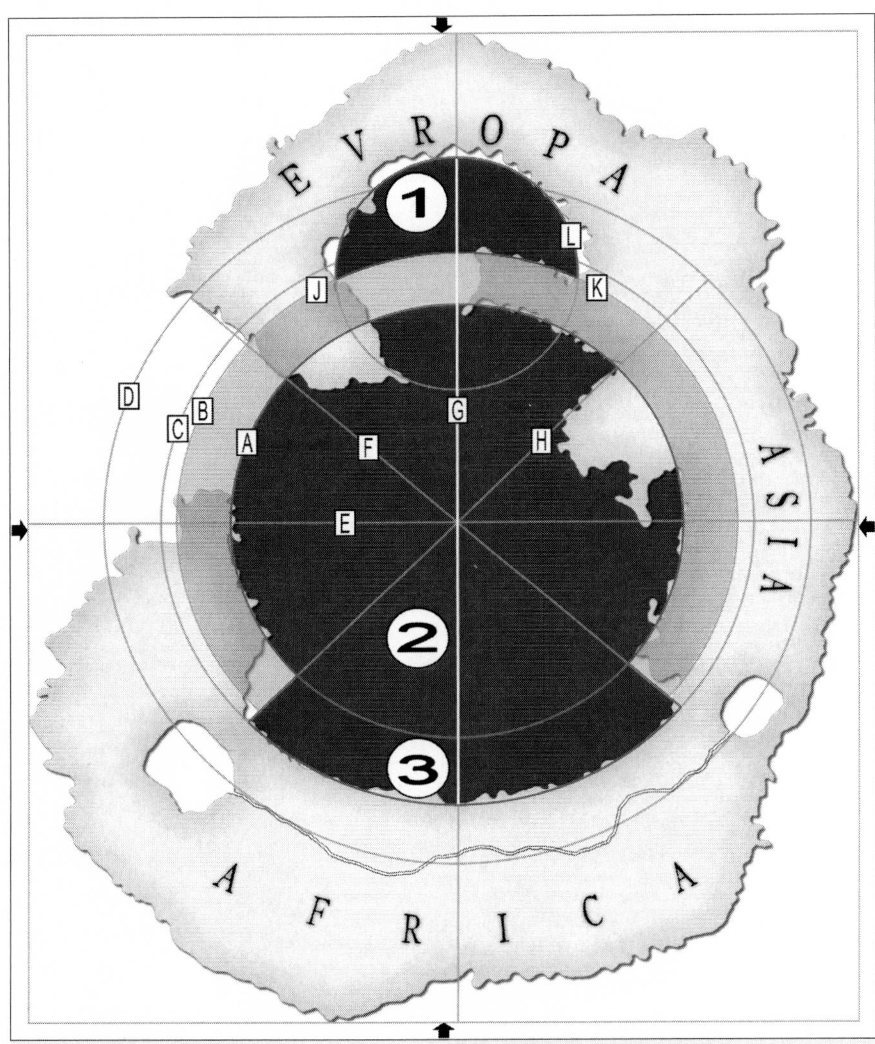

Figure 37. *Author's reconstruction of the map's original framework and orientation. In the process of creating the original map, a set of perpendicular lines, E and G, were drawn to set the four cardinal directions. Bisecting lines, F and H, established the western coasts of the Iberian and Greek peninsulas and the lateral extent of the lower North African coast. Concentric circles A through D, centered on the map, established guides for interior coastlines and the positioning of the Upper Nile. Circle L, centered on centerline G, sets the coastline west of Italy. The final design divides the Mediterranean into 3 geometrically aligned zones: 1, 2, and 3.*

north central Africa and the Middle East, and the last ring (D) dictates the path of the mythical meandering Upper Nile.

These rings were not only guides for coastlines and waterways; they also established a center point upon which was built the remaining framework. There are two lines that align with this center point in the form of two of the map's straightest stretches of coastline. These are the western coasts of the Iberian and Greek peninsulas, which also happen to be the only two significant intrusions into the map's innermost circle, Zone 2. Figure 37 draws two lines (F and H) running along these coasts and through the common center point of the concentric circles. The lines are perpendicular to each other and effectively divide the inner circle into four equal parts.

We can confirm that these lines were indeed part of the framework. The two lines intersect the lower North African coast that runs along circle C, roughly coinciding with the width of the African coast just before it rises at Mauritania in the west and Turkey in the east. This also establishes the second largest section of the Mediterranean, Zone 3, which arcs between Zone 2 and Africa's northern coast.

The third significant portion in the design of the Mediterranean, Zone 1, is located along Italy's western coast. In Figure 37, circle B delineates the lower surface of this space, while the upper half of circle L defines the arched coastline opposite Italy's western coast. Circle L has been created by drawing a circle through circle B's intersection with the European coast (points J and K), with a center point set equidistant to those intersections on centerline G.

With this framework established, the linear nature of the map becomes clear, the map defines the Mediterranean with three center-aligned geometric zones: Zones 1, 2, and 3. Utilizing the centerline running through the midpoints of these three zones, the map has been rotated so that the centerline runs vertically with the arched Zone 1 positioned toward the top. The reasoning behind this orientation is based on the shape of Zone 1. The zone is a semicircular shape mimicking the Roman arch. This architectural creation, popularized by the Romans, is oriented arching upward above structural openings, so it is fitting that the Romans purposely distorted the Iberian Peninsula in order to create this arched zone rising majestically above Italy and the city of Rome. Unlike the Greek and medieval maps that placed Greece and Jerusalem at their centers, here the Roman mapmaker places Rome at the top of the map.

One of the noticeable effects of orienting the map with the arch at the top of the map, which further supports this as the original orientation, is how it places the inner circle (Zone 2) very near the map's horizontal and vertical center. The four arrows along the map's perimeter mark the map's true horizontal and vertical centers, demonstrating the slight variance between Zone 2's center and the map's actual center. The sense of the map's original symmetry can also be seen in that the horizontal line drawn through the map's center (E) bisects Asia at its outermost point, with the coastline falling away on either side at a similar angle. Similarly, the vertical line drawn through the map's center (G) bisects Europe in the north at its highest point, with the coastline again falling away on either side at similar angles.

It may seem odd to refer to the alignment of the map as having a northern orientation while the Italian Peninsula is clearly lying on its side, but we determine the alignment based on the positioning of the continents: Europe located above Africa and Asia off to the right. This also proves true of the Roman Peutinger map, which also exhibits a clear northern orientation while similarly placing the Italian peninsula on its side.

The location of the upper arched zone above Rome suggests that this area was set aside to honor Rome or one of its rulers, much like the Peutinger Table features an illustration of a woman on a throne framed by a large inscribed circle at the spot where the Eternal City should be (Fig. 34).

Figure 38 resurrects Agrippa's *Orbis Terrarum* with a reconstruction based on Schöner's southern landmass. The three major zones composing the oversized rendering of the Mediterranean Sea can be seen accommodating components that would later be redistributed about the *mappae mundi* (Fig. 39). Zone 1, which is located at the top of the map adjacent to Rome is fitted with a banner inscribed with the name "Augustus." A semicircular frame rises above it. This arched frame houses an image of Caesar Augustus—bestowing honor and credit to the man responsible for commissioning the map. *Mappae mundi* like

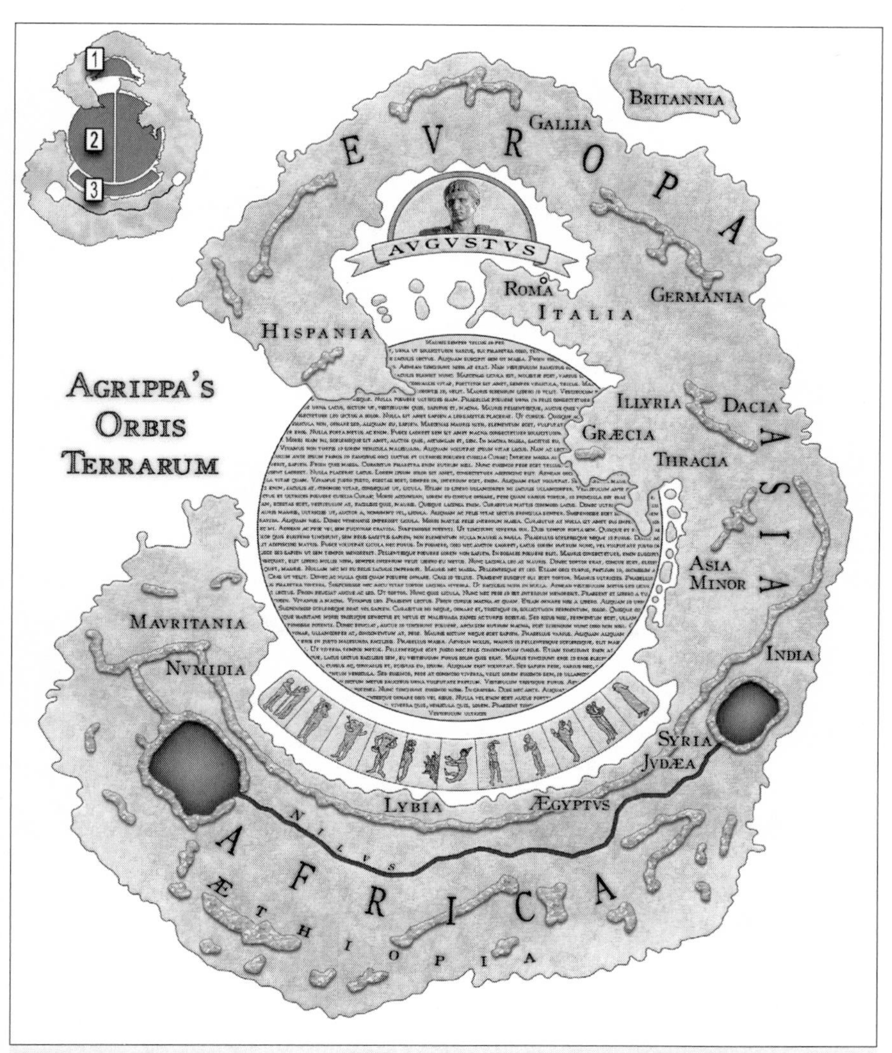

Figure 38. *Author's reconstruction of Agrippa's* Orbis Terrarum *based on the map at the bottom of Schöner's 1515 globe and its internal geometric guidelines. The reconstruction reflects a northern orientation, with Europe toward the top and Africa below, much like the Roman Tabula Peutingeriana. The three zones (see inset) are filled with: 1) a tribute to Caesar Augustus, the man responsible for commissioning the survey that would lead to the map's creation; 2) an extensive commentary on the geography of the world, as established through Agrippa's survey; and 3) a grid displaying a variety of creatures and plants believed to inhabit various regions of the known world.*

Figure 39. *Sections of the Ebstorf* mappa mundi *with insets of Agrippa's* Orbis Terrarum, *the map upon which the* mappae mundi *were based. Here you can see portions of the commentary that were originally consolidated in the center circle of Agrippa's map.*

The top section of the Ebstorf is emblazoned with an image of Christ. This would have been a natural adaptation by medieval Christians, likely replacing the image of Caesar Augustus atop Agrippa's map in the upper arch with an image of Jesus on the most prominent position on the mappae mundi.

Lining both sides of the arced Upper Nile (lower image) on most mappae mundi *are consolidated bands of flora, fauna, and humanoids. These, along with the rest of the flora and fauna distributed about the map, likely occupied the similarly downward arcing southernmost zone of Agrippa's map.*

All of these adjustments appear to have been necessitated by the medieval Christian desire to extend Jerusalem into the center of their designs, forcing Agrippa's centralized elements to be moved directly onto the map.

the Hereford and Psalter maps still retain this space but relocate it at or above the top of the map, replacing the Roman representation with an image of Jesus Christ.

Zone 2, the innermost circle, has been fitted with a lengthy commentary that would later be broken up and distributed throughout the mappae mundi according to its described native region. The amount of text on the original map can only be roughly approximated by a number of unknown factors, among them the map's overall size, text size, and line spacing. As a rough example, if we scale the map to the Hereford's 52" width and use modern standard rule spacing for lines of text, the total commentary fitting within this inner circle could exceed 2,000 words, or about seven standard pages of double-spaced text—a pretty reasonably sized commentary. If we apply the same ruled spacing but grant equal importance to Agrippa's map as was given the largest known *mappa mundi*, the Ebstorf (which measures nearly 12 feet in diameter), the commentary balloons to over 14,000 words, approaching 50 standard pages of text, which would allow for the inclusion of an astonishing wealth of geographic data.

Zone 3, the arcing band located between Turkey and Mauritania, had a fairly straightforward usage. This arced band did not completely disappear during the Christocentric transition but found much of itself migrating directly southward below the mythical Upper Nile. This feature, as found on the Hereford, Ebstorf, Psalter, and many other *mappae mundi*, still retains its downward arc and is segmented into multiple boxes framing images of the flora, fauna, and humanoids believed to have inhabited the African continent. Many of these creatures are documented in Pliny's *Natural Histories*; since Agrippa's map was constructed within that time period and influenced by a Roman mindset, it likely incorporated them into Zone 3. The medieval *mappae mundi* appear to have borrowed them directly from Agrippa's map, even retaining the zone's relationship to the Nile, being both parallel and constrained closely to the length of the waterway.

The reconstruction borrows back the chain of creatures located beneath the Hereford Map's Upper Nile and returns it to its original location in Zone 3. In comparison to the depiction, the original arced chain was likely a grid comprising more columns, rows, or both, allowing it to accommodate representations of all the map's flora, fauna, and an array of imagined creatures like the monoculi, single-footed humanoids that used their giant foot to shield themselves from the sun, and the blemmyae, war-like creatures with their faces embedded in their chests.

While many of the *mappae mundi* retained intact a sizable section of the original grid in Africa, the world's remaining plants and creatures were detached from the grid and dispersed throughout Europe, Asia, and other regions of Africa according to native region. Due to Agrippa's centralized design, each creature would have to have been accompanied not only by an inscription providing the specific name and description, as is done on the *mappae mundi*, but also text specifying each creature's native land. Agrippa's overall design is reminiscent of a modern atlas, having a map depicting the standard cartographic detailing of mountains, waterways, towns, cities, and perhaps roadways, while a separate array of accompanying notes and imagery provided viewers with in-depth information beyond the normal cartographic features; details of plants, creatures, and terrain a traveler might experience on the ground should they travel to the specific regions plotted on the map.

With the authenticity of the source map established as a copy of a Roman world map, most likely Agrippa's *Orbis Terrarum,* and having had a chance to conceptualize its original design, we also establish a key aspect of Step 1 in Schöner's process. Like Piri Re'is did in creating his 1513 map, Schöner was indeed referencing ancient maps for his 1515 globe, demonstrating a preference for relying on these ancient sources of information rather than creative license when depicting his Antarctic continent.

Schöner's source map was most likely a version of Agrippa's world map that never made it to completion. Maps went through multiple stages of construction; the design was first sketched out in full, land would be painted in next, followed by a layer of mountains, then waterways, and finally text would be added and fitted around the terrain. Schöner's source map likely only existed in incomplete outline form as his landmass lacks major inland bodies of water like the Black and Red Seas, which had been basic features of maps for centuries. Because of its incomplete nature, it was likely relegated to a specific section of a library, set aside for incomplete works similar to canvas portraits that were sketched but never painted. Whatever the case may be, it is certain that Schöner had access to a few such maps and was open to the idea that contemporary discoveries may in fact be rediscoveries of lands once known and charted but long forgotten. And if there were compelling enough evidence that ancient depictions matched up convincingly to new geographic finds, what need was there to reinvent new designs? This brings us to Step 2 of Schöner's methodology.

Step 2: Reconciling Discoveries to Ancient Source Maps

First, recall what instigated Schöner's insertion of this Roman world map onto the bottom of his 1515 globe. Again, it was not the discovery of the Strait of Magellan, but rather the discovery and partial exploration of the San Matias Gulf, which sailors returning from South America had erroneously purported to be a strait very much like that of Gibraltar. The overall account lends itself to the visualization of a large continental landmass like Africa separated from a point of land like Spain's Gibraltar by a strait of water.

> *They have sailed around that point, and ascertained that the country lay, as in the south of Europe, entirely from east to west. It is as if one crossed the Strait of Gibraltar to go east in ranging the coast of Barbary.*

Provided this account, how could Schöner help but envision a long coastline similar to the Barbary or the North African coast separated from a point of land—the tip of South America—by a strait? Armed with this imagery, we can see Schöner rummaging through a collection of ancient maps in hopes of finding features that fit the sailors' description.

Schöner, who obviously had experience with globes prior to designing his 1515 globe, would have immediately recognized what he was seeing when he came across this unfinished map of a landmass with its concentric guidelines still exposed. There is only one location where a landmass could be delineated in this fashion: on the top or bottom of a globe, where the rings would represent latitudes and the lines intersecting the center of the concentric rings would represent longitudes.

Having believed he had discovered a map of a polar landform, all that was left to determine was the location of the sailor's purported strait. If this were a legitimate Roman map, there would have existed an obvious strait ready to fit this requirement: the English Channel. Schöner seems to have recognized this correlation as we see on his 1515 globe, where the tip of South America is pocketed into the side of the accommodating, curving coastline of his southern landmass. Britain's southeastern coast similarly protrudes out toward a conforming European coastline.

Step 3: Scaling Old to New

Schöner now had two recognizable points on his copy of Agrippa's map: a polar point at the center of the map's concentric framework and a strait between a large unknown landmass and a small point of land.

The map may have been worn along the outer edges, obscuring the fact that the point of land was actually part of an island and not a larger landmass. Even if it had been clearly drawn as an island, the large landmass on a polar grid likely would have been irresistible. Besides, Schöner had already witnessed the Americas being drawn initially as small defined islands, only to see them enlarged on subsequent maps after further exploration (Fig. 14). He, in fact, had made the same error himself. It would therefore not be a huge leap of logic for him to assume that this British Isle on Agrippa's world map was similarly an early ancient rendering of the tip of the South American continent.

At that point, Schöner was prepared to scale the ancient map to his new globe. The process entailed placing the center of the concentric rings at the southern pole and stretching the English Channel up to the 40th parallel, where the Portuguese sailors had implied the existence of a strait lying between the two landmasses (Fig. 40). The British Isle now melded into South America as the continent's southern tip. This brings us to the final resting place for the last known remaining copy of an ancient Roman world map.

In summary, there is at least one thing of which I am certain: Schöner

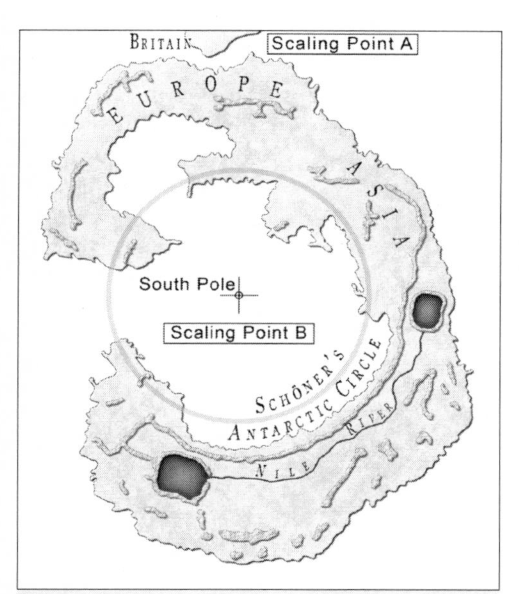

Figure 40. *Schöner's scaling of an ancient map of the world onto his Globe of 1515. At the top, the first scaling point is the English Channel, which sits in place of a purported strait passing from the Atlantic to the Pacific. The second scaling point is the center of Agrippa's circular commentary delineated by the concentric coastlines of Italy, Turkey, and Mauritania, which he centers over the South Pole.*

did indeed preserve a copy of Agrippa's *Orbis Terrarum* on his 1515 globe. But I am not fully convinced—nor, for that matter, should anyone be at this point—that his 1524 globe was based on a genuine map of Antarctica. There are far too many questions this assumption leaves unanswered. Where did the source map originate? Who charted the continent and when? How did the map find its way into Europe? How and why was it preserved until the sixteenth century? And of course, based on the map's ice-free depiction of the continent, how could what is believed to be millions of years of Antarctic ice accumulation have occurred within a few thousand years?

Yet the apparent shared methodology between the two globes (Fig. 41) should be afforded serious consideration, as Schöner's incorporation of Agrippa's Orbis Terrarum onto his 1515 world globe increases exponentially the likelihood that a deglaciated Antarctica was visited and charted in the ancient past. The complexity of the elements within Schöner's 1524 depiction that match Antarctica is compelling in itself. And since his incorporation of Agrippa's map establishes a clear methodology with his 1515 globe of reconciling and scaling ancient maps to new discoveries, this increases the probability of adherence to a similar methodology being used on his 1524 globe. In this case, there is absolutely no better geographically matching candidate for the source map than a map of a deglaciated Antarctica. All in all, this may not be

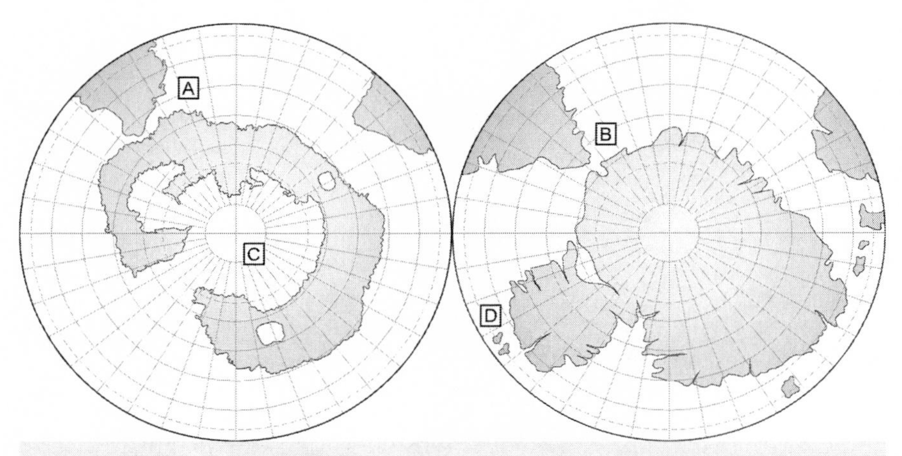

Figure 41. *Schöner's Methodology for Cartographic Incorporation of New Discoveries:*
Step 1. Referencing ancient maps for his template: Agrippa's Orbis Terrarum in 1515 (left) and an ancient map of Antarctica in 1524 (right),
Step 2. Reconciling the ancient maps to new discoveries: (A) Matching the English Channel to a purported strait and (B) Atka Bay to a southern bay in the Strait of Magellan, and
Step 3. Scaling the maps to new globes via a secondary point: (C) aligning the center of a concentric Mediterranean to the South Pole and (D) the islands of Carney and Siple to the Unfortunate Islands high in the Pacific.

enough to overturn established history—or scientific dating methods, for that matter—but it is a strong enough argument to significantly expand Hapgood's initial crack in Antarctica's 34-million year old ice cap.

Personally, I find it hard to fathom that this is all mere coincidence. It is eerily reminiscent of one of those movie plots where the main character is led to a specific destination or exposed to hidden truths by a mysterious acquaintance only to find out in the end that his guide never existed. In this instance, an ancient map of Antarctica, which may or may not be genuine, led first to the discovery of an obscure, little-known island set hidden under an ice shelf off the coast of Western Antarctica. Then the map's perceived method of two-point scaling led to the discovery of an ancient Roman map placed on the bottom of another globe produced by Schöner because it utilized the same method of two-point scaling. It leaves me asking: If the end result is genuine, could the mysterious guiding force—an alleged ancient map of Antarctica—have been real as well?

CHAPTER 6

IN SEARCH OF ATLANTIS

Regardless of how strong the evidence may appear, it remains a daunting and improbable task to identify an ancient civilization capable of not only sailing to the remote Antarctic continent but circumnavigating and charting it in such detail.

It seems certain that the map was not charted by Europeans of Schöner's day. There was some secrecy maintained around discoveries of the time due to fear of rival countries sharing or claiming rights to regions that were a source of wealth, but it did not stop the proliferation of maps and globes depicting the Americas. Additionally, Schöner's 1524 Antarctic design is a highly accurate map of the continent—far more accurate than maps of the Americas being produced at the turn of the sixteenth century. As would be expected, European exploration saw South America evolving steadily into more accurate depictions, but the Antarctic maps were most accurate when they first appeared on maps and globes of the early sixteenth century. After this, the southern continent quickly devolved into a completely unnatural and unrecognizable landmass as exploration further southward found only more ocean. In response, cartographers began slowly cutting away at the overscaled version of the continent. The sudden insertion of an accurate but misplaced and oversized Antarctic continent by Schöner means that he was not getting the information through coordinates provided by European explorers as was common with the Americas. Therefore, the continental design predated sixteenth-century Europeans who did not grasp the map's true scale.

There is an ancient record with detailed evidence of a civilization possessing the navigational skill required to make the journey, but like the ancient map of Antarctica, the likelihood of its existence remains in doubt among modern historians. That civilization is said to have existed far beyond the Mediterranean—in a land called Atlantis.

If Atlantis truly existed, then we should be able to make a reasonable assessment of its current location. Contrary to popular belief, this is not a needle-in-a-haystack proposition. The biggest public misconceptions regarding Atlantis relate to its size and location. An inside-the-box approach among historians, anthropologists, and archaeologists, which clings to a shared established view of our past, restricts the geographic range and technological advancement of ancient seafaring civilizations. This has necessitated enormous disregard for the detailed description of Atlantis contained within two of Plato's dialogues, *Critias* and *Timaeus*.

Some choose to dismiss the Atlantis tales within these dialogues altogether, claiming they are works of fiction or allegory meant to reflect the ancient Greek philosopher Plato's view of the perfect society. While that is certainly possible, it is worth noting the contrast in Plato's presentation of allegory versus factual (or at least perceived factual) events. In Book VII of one of his most important works, *The Republic*, Plato presents a dialogue having occurred between his brother Glaucon and his mentor Socrates:

> *Socrates: This entire allegory, I said, you may now append, dear Glaucon, to the previous argument ... Which desire of theirs is very natural, if our allegory may be trusted.*

The above passage is from a work known as *The Allegory of the Cave*. Here Socrates openly confirms that his story of men chained in a cave from birth is allegory, a discussion of fictional characters and events he is using to illustrate "how far our nature is enlightened or unenlightened." Contrast this with the presentation of Atlantis:

> *Critias: Then listen, Socrates, to a tale which, though strange, is*

certainly true, having been attested by Solon, who was the wisest of the seven sages.

Socrates: Very good. And what is this ancient famous action of the Athenians, which Critias declared, on the authority of Solon, to be not a mere legend, but an actual fact?

Critias: "Tell us from the beginning," said Amynander, "what Solon related and how, and who were the informants who vouched for its truth." [7]

The presentation of the Atlantis event that Critias goes on to relate was intended to provide Socrates with an example of the ideal society. This could easily have been achieved through allegory with all the impact and conceptual conveyance attained in Plato's *Allegory of the Cave*. Yet where we see repeated assurances in the *Allegory of the Cave* that the tale is indeed allegory, here in Timaeus we find multiple assurances that the Atlantis tale, "though strange" in light of known history and the worldview of Socrates and Critias, was indeed true.

The story is presented as fact. Whether Plato was being intentionally deceptive in passing off a complete fiction of his own making as fact is another matter to debate. But the alleged truth is that the Athenian statesman and poet Solon visited the district of Sais in Egypt where an Egyptian priest conveyed to him a history and extended geographic view that was completely foreign to him and the people of Greece.

In the Egyptian Delta, at the head of which the river Nile divides, there is a certain district which is called the district of Sais, and the great city of the district is also called Sais.

To this city came Solon, and was received there with great honour; he asked the priests who were most skillful in such matters, about antiquity, and made the discovery that neither he nor any other Hellene knew anything worth mentioning about the times of old.

One of the priests, who was of a very great age, said: O Solon,

---

[7] *Timaeus* 20-21. *Timaeus* by Plato; translation by Benjamin Jowett utilizing Stephanus pagination. Unless otherwise noted, all quotes of Plato's Timaeus and Critias are from the Benjamin Jowett translation.

Solon, you Hellenes are never anything but children, and there is not an old man among you.

You do not know that there formerly dwelt in your land the fairest and noblest race of men which ever lived, and that you and your whole city are descended from a small seed or remnant of them which survived.

Solon marveled at his words, and earnestly requested the priests to inform him exactly and in order about these former citizens.[8]

This brings us to the first level of acceptance by mainstream historians. Wishing to believe that the story has its roots in an actual historical event, like many legends, some find it necessary to reconcile the Atlantis account within a constrained framework. While locations for Atlantis have been proposed all over the world, the most widely-accepted theory is that the legend grossly exaggerated a cataclysmic event that occurred within the Mediterranean. This limitation has seen the ruins of Santorini and Crete promoted as locations for Atlantis. In truth, however, these locations can never be linked positively to Atlantis. While the sites comply with Plato's account of a cataclysmic event having devastated the land, they fail to meet the bulk of criteria Plato provides about the geographic size, location, and layout of Atlantis.

When we adhere more closely to Solon's detailed description in Plato's dialogue, we can eliminate these two sites and any other sites proposed in the Mediterranean.

A mighty power which unprovoked made an expedition against the whole of Europe and Asia. This power came forth out of the Atlantic Ocean [from] an island situated in front of the straits which are by you called the Pillars of Heracles.[9]

Solon could hardly be clearer. Not only does he locate Atlantis beyond the Pillars of Hercules—an ancient Greek term for the Strait of Gibraltar—but he also names its location as the Atlantic Ocean outright. Some have persisted in ignoring Solon's Atlantic location, theorizing that the Pillars of Hercules originally referred to a different strait within

---

[8] *Timaeus* 21e-23d

[9] Ibid., 24e

the Mediterranean to bolster the credibility of such sites as the islands of Sardinia and Malta, but Solon introduces another key element that further negates this possibility. Solon declares Atlantis to have been a large continent-sized landmass:

> *The island was larger than Libya and Asia put together.*[10]

While the Greeks had limited and varying perceptions of the size of Asia and Libya, a term then applied to the African continent, they viewed the two combined equal to or larger than Europe. Hence, Atlantis was perceived as an island larger than the combined continents of Asia and Libya, and by extended logic, both larger than Europe and distinct and separate from the three continents as well. This narrows our search for Atlantis to a continent-sized landmass beyond the confines of the Mediterranean and lends support for an Atlantic location.

One popular hypothesis holds that Atlantis still lies beneath the watery depths of the Atlantic. This is based on the following passage in Plato's Timaeus:

> *There was an island situated in front of the straits ... In a single day and night of misfortune all your warlike men in a body sank into the earth, and the island of Atlantis in like manner disappeared in the depths of the sea.*[11]

In his 1664 work titled *Mundus Subterraneus*, Athanasius Kircher, a German Jesuit scholar, documented his support of this theory of a large sunken continent and incorporated a map depicting Atlantis' positioned in the middle of the Atlantic Ocean between Europe and the Americas (Fig. 42). An inscription in the top left corner makes clear the island's current location: "Site of the island Atlantis now beneath the sea, according to the beliefs of the Egyptians and the description of Plato."

The suggestion of a large submerged continent in the middle of the Atlantic is exceedingly problematic. Research aided by technological advancements throughout the twentieth century has allowed us to

[10] *Timaeus* 24e
[11] Ibid., 24e, 25d

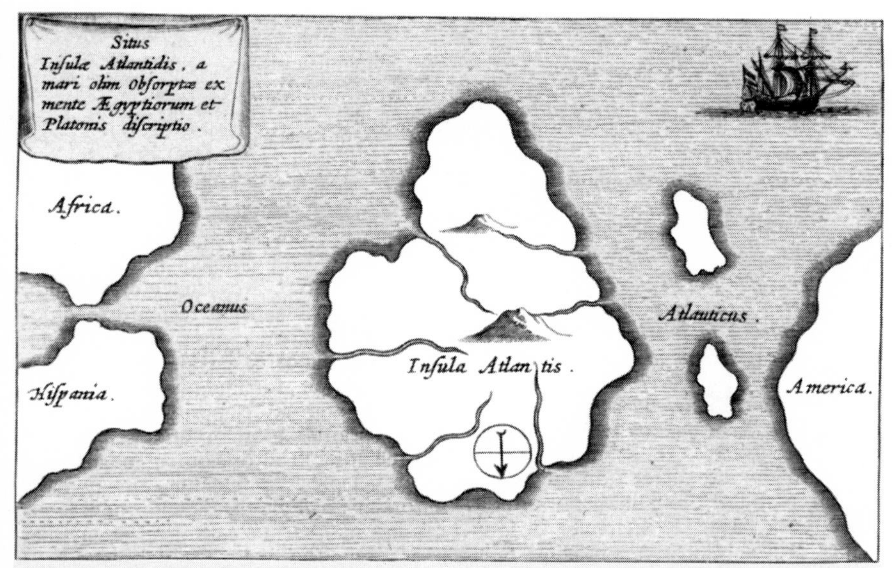

Figure 42. *A southern oriented map of Atlantis from Athanasius Kircher's* Mundus Subterraneus. *The map places the lost continent in the middle of the Atlantic Ocean between Europe and the Americas.*

chart highly detailed maps of the Atlantic seafloor. What we have found is a pattern of uninterrupted seafloor crustal expansion extending away from a central longitudinal expansion ridge known as the Mid-Atlantic Ridge (Fig. 43). There are no excessively large sunken plateaus or evidence of an undersea continental mass anywhere in the Atlantic and only a few relatively small plateaus associated with undersea ridges. Additionally, there are just a few small islands, including the Azores, which originated from ruptures in the oceanic crust. In fact there are no continent-sized landmasses approach-

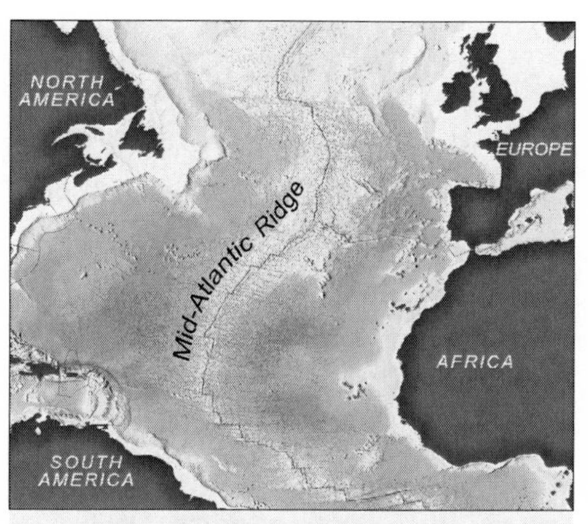

Figure 43. *Bathymetric map of the Atlantic seafloor, which is devoid of sunken continents.*

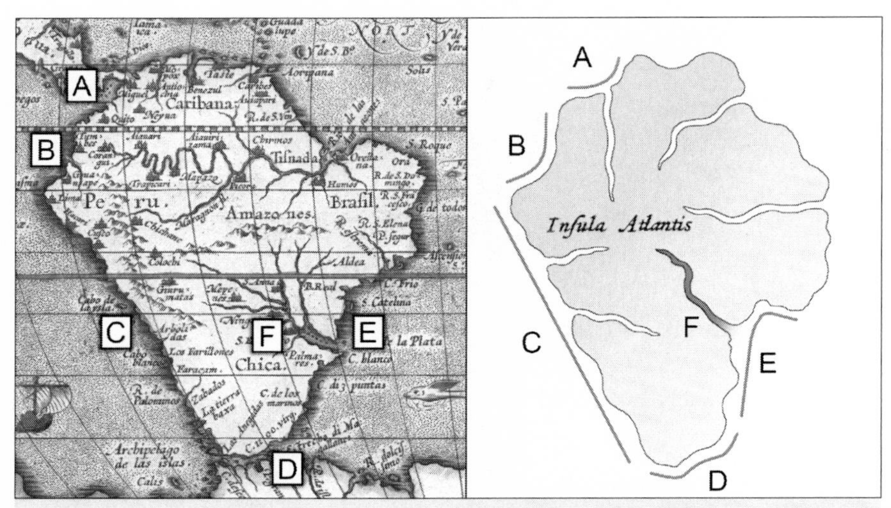

Figure 44. 1592 Typus Orbis Terrarum *by Abraham Ortelius (left), which shares an uncanny likeness with Kircher's Atlantis (right, reoriented with north toward the top). They correctly depict a recessed southeastern coastline interrupted by the Rio de la Plata (E,F), but both also present erroneous depictions of a straight western coastline (C) as well as a blunt west to east rising southern tip (D). Both also depict an almost identical transition to a double scalloped coast in the northwest (A, B).*

ing the size of Europe, Asia, or Africa that exist submerged beneath any of Earth's oceans. This restricts our identification of Atlantis to a limited few continents currently breaching the oceans' surface.

There is an aspect of Kircher's map that I found interesting: his design of the Atlantis continent. The design appears to have been borrowed from maps of the South American continent produced decades earlier. The overall shape of the continent is virtually identical to South America as depicted on the 1592 *Typus Orbis Terrarum* by Abraham Ortelius (Fig. 44). Both maps correctly depict a recessed southeastern coastline interrupted by the Rio de la Plata but incorporate the same erroneous depiction of a perfectly straight and slanting western coastline as well as a blunt west-to-east rising southern tip. Kircher's Atlantis even precisely mimics the southward pointing orientation found on Ortelius' version of South America, disregarding the longitudinal delineations that accurately convey South America's southwesterly orientation.

Although tempted by the possibility that Kircher's map was based on an ancient map of South America, there is no clear evidence to substantiate it. To be a worthwhile consideration, Kircher's South American design would ideally have appeared on maps charted before or nearer to 1492 in order to negate contemporary exploration as the true source and inspiration of the design. As it stands, there are a string of maps beginning at the turn of the sixteenth century that show a continually evolving portrayal of the South American continent, morphing over time to its current iteration with more and more extensive exploration of the region.

Still, questions remain as to why Kircher would depict Atlantis with a map that resembles a decades-old rendering of South America. At the time of the map's creation, there were books written by Sir Thomas More and Sir Francis Bacon, which—although fiction—linked the Americas to a Utopian society like Atlantis. If this influenced Kircher to depict Atlantis as South America, it makes little sense that he would replicate the continent and locate it in the middle of the Atlantic. Perhaps he was generating a random design, but if so, why does it so accurately mimic maps of the continent that originated seven decades earlier? I believe the most reasonable possibility is that he, similar to Piri Re'is and Schöner, was referencing an older source map and scaling an actual map of South America to the middle of the Atlantic. The source map, however, was not one charted by an ancient Atlantean civilization, but one that originated only decades earlier. Still, one has to wonder, could he have had such little experience with maps that he did not recognize the design's similarities to contemporary portrayals of South America? We will never know. I admit to remaining baffled by Kircher's design. However, it is hard to dismiss it entirely, especially considering what is about to unfold as we continue in our search for the fabled land.

As I began my own investigation into the site of Atlantis, I decided that it was best to identify a geographic match based on the highly detailed description Plato's dialogues provided and adhere to it as closely as possible. This was as close to the source as we can currently get, so I started there and moved forth systematically. The most logical starting point seemed to be identifying a large geographic feature the account claimed existed in the midst of Atlantis—a feature with distinct attributes that would make it the central focus of my search. That feature is an extremely large, uniquely rectangular plain. As it turned out, Plato had included a component that would prove an invaluable aid in narrowing down the search: the feature's dimensions.

> *The whole country was said by him to be very lofty and precipitous on the side of the sea, but the country immediately about and surrounding the city was a level plain, itself surrounded by mountains which descended towards the sea; it was smooth and even, and of an oblong shape, extending in one direction three thousand stadia, but across the centre inland it was two thousand stadia.*[12]

> *It was originally a quadrangle, rectilinear for the most part, and elongated; and what it lacked of this shape they made right by means of a trench dug round about it ... and since it was dug round the whole plain its consequent length was 10,000 stades.*[13]

Yet I was not the first to recognize the potential of this uniquely laid out plain in helping identify the lost continent. For many decades, the best geographic match for Atlantis' unique plain was offered by another Atlantis theorist, Jim Allen. In the 70s, Allen had set out on his own search for the Atlantis plain and ultimately linked it to the Altiplano (Fig. 45), a large plain located high up in the Andes Mountains of South America in the western region of Bolivia. The plain possesses the requisite rectangular shape and is clearly "surrounded by mountains," so it matches the most basic aspects of the description.

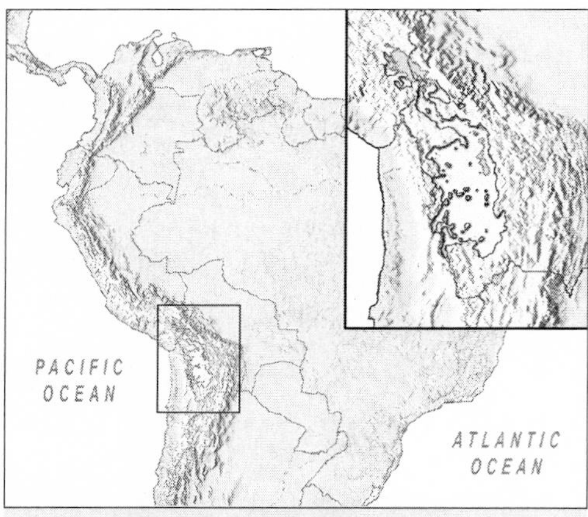

Figure 45. *Highlighted in the inset is the Altiplano, a large flat rectangular plain high up in the Andes Mountain range. British cartographer Jim Allen contends that this is the Atlantis plain and South America the island continent of Atlantis.*

[12] *Critias* 118a
[13] *Ibid.*, 118c, d; translation by R. G. Bury

And more importantly, Allen had boldly moved beyond the bounds of established history that favored Atlantis' location in the Mediterranean.

Allen's Atlantis hypothesis has sat at the forefront of geographic accuracy for quite some time, setting the bar for others like myself to clear. The following critique of the Altiplano site reveals where that bar actually sits and provides perspective on the challenges involved in locating and identifying the ancient plain of Atlantis.

Because the Bolivian Altiplano is substantially smaller than the Atlantis plain, Allen posits that the stadium once existed at roughly half the length of the ancient Greek and Egyptian stadium. Aside from the necessity of introducing a new measurement for the stadium, the proposed site also necessitates the introduction of a very large mountain lake, Lake Poopó, as a third sea. However, Plato's account makes absolutely no effort to distinguish or suggest any seas beyond two, the Mediterranean and Atlantic. This adjustment in interpretation was required because the original account places the sea very near the plain. A hill which would be the site of the capital city having a diameter of 27 stadia (3.1 miles), was said to lie 50 stadia (5.7 miles) from both the plain and the sea, combined measurements which require the plain to lie within 14.5 miles of the sea, or roughly 7 miles when applying the half-stadium. The only body of water lying within this range was the lake.

This further complicates issues as Plato's account claims that 1,200 naval ships manned by 200-man crews sourced from inhabitants of the plain navigated this sea and could sail via man-made channels from the sea, Lake Poopó, to the center of the city. Logistically we have to wonder why 1,200 ships this large and heavily manned would patrol the isolated waters of Lake Poopó, but even far more puzzling is how these men and their oceangoing ships ever made it down from their high perch in the Andes Mountains to engage the inhabitants of the Mediterranean.

One last obstacle pertains to the surrounding mountains. The account states that the mountains surrounding and sheltering the plain opened up toward the sea in the south, but the Altiplano is enclosed by mountains all around, and the proposed sea, Lake Poopó, is located in the east rather than the south.

This simple critique of the Altiplano hypothesis should demonstrate the extreme difficulty of identifying and associating a real-world plain as the Atlantis plain. The moment we begin comparing descriptions to the real world, discrepancies, small and large, arise. It is not so surprising given the extensive set of restrictive parameters provided by Plato's account, comprising a complex matrix of geographic features with relative alignments to each other, alignments to the cardinal directions, and finally dimensions, which are the most critical and extremely limiting of qualifiers.

On the upside, due to the presence of such intricate detailing, the account's description should prove extremely effective in locating and identifying the ancient plain and the long-sought continent of Atlantis if it truly existed. For all intents and purposes, Plato's account is a map in written form.

The Atlantis Plain

When I began analyzing the descriptive layout of the rectangular plain, the surrounding mountains were clearly significant, but I realized that they did not necessarily define the shape of the plain. The plain was defined by a navigable waterway; I felt that in the search for Atlantis, this rectangular waterway should be the key identifier.

While many people are aware of Atlantis' portrayal as a maritime power with an economy based in sea trade, many may be unaware that it is also portrayed as having a robust river-based trade. Like many other ancient cultures, navigable waterways facilitated efficient transportation of the region's wide range of resources. The mountains' supplies of wood and precious metals as well as the plain's harvests, were transported down around the plain's perimeter. They would travel around it and arrive at the capital city, near where the surrounding waters converged and emptied into the sea.

> *The island itself provided most of what was required by them for the uses of life. In the first place, they dug out of the earth whatever was to be found there, solid as well as fusile, and that which is now only a name and was then something more than a name, orichalcum, was dug out of the earth in many parts of the island,*

being more precious in those days than anything except gold. There was an abundance of wood for carpenter's work, and sufficient maintenance for tame and wild animals.[14]

I will now describe the plain, as it was fashioned by nature and by the labours of many generations of kings through long ages. It was for the most part rectangular and oblong, and where falling out of the straight line followed the circular ditch. The depth, and width, and length of this ditch were incredible, and gave the impression that a work of such extent, in addition to so many others, could never have been artificial. Nevertheless I must say what I was told. It was excavated to the depth of a hundred feet, and its breadth was a stadium everywhere; it was carried round the whole of the plain, and was ten thousand stadia in length. It received the streams which came down from the mountains, and winding round the plain and meeting at the city, was there let off into the sea. Further inland, likewise, straight canals of a hundred feet in width were cut from it through the plain, and again let off into the ditch leading to the sea: these canals were at intervals of a hundred stadia, and by them they brought down the wood from the mountains to the city, and conveyed the fruits of the earth in ships, cutting transverse passages from one canal into another, and to the city.[15]

This portrays Atlantis as a self-sustaining riverine culture rivaling the likes of other ancient civilizations like the Babylonians and Assyrians on the Euphrates and Tigris Rivers and the ancient Egyptian, Chinese, and Indian cultures based on the Nile, Yellow, and Indus Rivers, respectively. Maintaining these fully navigable waterways requires a great deal of runoff from the surrounding mountains. Logically, if all these other great ancient waterways still flow and remain navigable, we should expect that the extensive waterway of the Atlantis plain should still exist in some form today as well. Figure 46 provides a sense of how large the Atlantis plain and its rectangular waterway were in comparison to the Mesopotamian plain lying between the Euphrates and Tigris Rivers.

---

[14] *Critias* 114e

[15] Ibid., 118c-118e

Figure 46. *A 2,000 x 3,000-stadium rectangle representing the Atlantean plain, scaled and overlaid onto an image of Mesopotamia. The Atlantean plain, with a border defined by a navigable waterway and a multitude of irrigation channels dug through it would have rivaled other ancient riverine civilizations of the past whose waterways still exist and are in full use today. Logically, if these other great waterways still exist, we should expect that the Atlantis waterway should still exist in some form as well.*

It is certainly possible that the waterway was altered over thousands of years, but I highly doubt that such a substantial waterway would have completely dried up and stopped flowing from the surrounding mountains. In fact, based on the purported scale of the surrounding mountains as the source of these waters, mountains that according to the account were "celebrated for their number and size and beauty, far beyond any which still exist," one would expect meltwater to flow down in similar abundance today.

I began searching the globe for a large waterway that retained a rectangular form. After a very thorough search, it became apparent that there was only one such plain in existence—and it happened to be almost precisely the size specified for the Atlantis plain. The plain lies in Argentina, South America, and is defined by two surrounding rivers, the Paraná and Uruguay. Like its namesake in the Middle East, it is called Mesopotamia (Fig. 47), the "land between rivers."

The Atlantis plain was said to measure 10,000 stadia, or roughly 1,150 miles around its rectangular perimeter. How well does this plain match up? The Paraná and Uruguay Rivers drop out from the Brazilian Highlands and onto the Argentine plain in the north, landing some distance apart. Here, the foothills between the two rivers effectively seal off and define a 70-mile portion of the plain's northern border. From this point the Uruguay River flows 410 miles southward, forming the

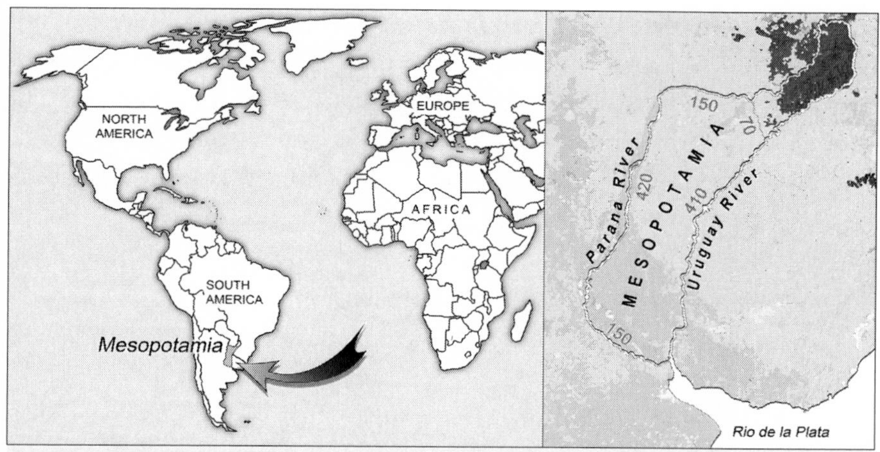

Figure 47. *The Mesopotamian plain, located in northern Argentina, is the world's largest rectangular plain that is defined by waterways that match Solon's description. A topographic map of the Mesopotamian plain (right) with its 1,200-mile (10,455 stadia) perimeter defined by the Paraná and Uruguay Rivers and the Brazilian Highlands to the north.*

eastern edge of the plain before eventually converging with the Paraná River in the south. Where the Paraná River drops onto the plain in the north, it immediately veers westward roughly 150 miles, forming the remainder of the northern border, then drops 420 miles southward, forming the plain's western border. The Paraná finally veers to the southeast 150 miles to meet up with the Uruguay River, fully enclosing the plain at its southern end and defining the Mesopotamian plain's rectangular perimeter. All these measurements combine to form a 1,200-mile perimeter, placing the Mesopotamian plain within a mere 50 miles of matching the size of the Atlantis plain.

Perhaps Plato was the beneficiary of extraordinarily good luck to land an estimate within 50 miles of the only plain that even remotely matches the size and description of the Atlantis plain. However, there are absolutely no other plains on the planet defined on all four sides by waterways that are larger, and none even approaching half the size of the South American Mesopotamia. Still, how far should we expect Plato's luck to extend? With only one plain in the world conforming to the size of the Atlantis plain, the odds of further conformity are not that good for a place that many believe is pure fiction.

Incredibly, the Mesopotamian plain's level of conformity extends well beyond size and shape. A topographic map of the region reveals that its entire layout conforms very closely to the Atlantis plain. The plain accurately receives the waters that flow down from the mountains surrounding it, and after the waters flow around the perimeter of the rectangular plain, they converge and empty out into the sea.

> *[The ditch] was carried round the whole of the plain, and was ten thousand stadia in length. It received the streams which came down from the mountains, and winding round the plain and meeting at the city, was there let off into the sea.*[16]

More specifically, the rivers converge and empty into the sea in the south. As can be seen very clearly in Figure 48, not only does the entire mountain-bounded region—which includes Mesopotamia and the extended plains beyond—open up or look toward the south, but it also accurately looks toward the sea as Plato described.

Figure 48. *Compelling evidence of the Atlantis site. This topographic map illustrates how perfectly the Mesopotamian plain matches up to the Atlantis plain. The 10,000-stadium rectangular plain defined by the Paraná and Uruguay Rivers not only meets size criteria, but the plain's perimeter is fed by tributaries originating from the surrounding mountains in the north, west, and east. The entire mountain-bound region accurately faces and is open toward the sea in the south where, as the account states, both waterways converge before emptying into the sea. There is no other site on the planet that approaches this level of conformance.*

> *The country immediately about and surrounding the city was a level plain, itself surrounded by mountains which descended towards the sea. This part of the*

[16] *Critias* 118c, d

> *island looked towards the south, and was sheltered from the north.*[17]

> *Looking towards the sea, but in the centre of the whole island, there was a plain which is said to have been the fairest of all plains and very fertile.*[18]

Unlike the Altiplano, which is completely surrounded by mountains and locates a small lake in the east in place of the sea, mountains surround Argentina's rectangular Mesopotamian plain on three sides and the fourth opens to a true sea in the south, the Atlantic—a sea that would allow the occupants of the plain to sail ocean-worthy vessels to the Mediterranean.

There are two more interesting aspects to the location of the Mesopotamian plain. The first is its climate: its location near the tropics makes it conducive to biannual harvests providing the Mesopotamian plain with a favorable agricultural advantage. This rich harvest is also ascribed the plain of Atlantis.

> *Twice in the year they gathered the fruits of the earth—in winter having the benefit of the rains of heaven, and in summer the water which the land supplied by introducing streams from the canals.*[19]

The other intriguing aspect of the plain's location relates to Plato's tale as told from the perspective of someone having traveled from the Mediterranean. It is normal for a traveler to describe a location as he saw it during his approach. In fact, Critias, in attempting to convey Solon's words, prefaces his description of the whole of Atlantis by stating that he will present its arrangement or order.

> *I have described the city and the environs of the ancient palace nearly in the words of Solon, and now I must endeavour to represent the nature and arrangement of the rest of the land. The whole country was said by him to be very lofty and precipitous*

---

[17] *Critias* 118a, b
[18] Ibid., 113c
[19] Ibid., 118e

on the side of the sea, but the country immediately about and surrounding the city was a level plain.[20]

Ferdinand Magellan stayed near the western coast of Africa as far as the Cape Verde Islands, at which point he set a course south-southwest across the Atlantic for Brazil. After traversing the Atlantic and making landfall along the eastern coast of South America at Cape St. Augustine, Brazil, the journey southward toward Rio de la Plata was set along a lofty and precipitous coastline. The Brazilian Highlands form nearly half of Brazil's topography and extend out to the eastern coast, creating a precipitous backdrop for a few scattered slivers of low-lying coastal regions. Approaching the Rio de la Plata midway down the coast, the terrain does indeed transform dramatically into a large level plain.

If Atlantis had existed and its inhabitants traveled a similar route then conveyed a description of the region to the Egyptians as Plato's account states, they would likely produce a description similar to Solon's. The speaker would attempt to provide a vision that conveyed his experience of having traveled to his city, providing the listener with a visual of arriving at "a country very lofty and precipitous along the coast, but with the country immediately about and surrounding the city (in the vicinity of the Rio de la Plata) being a level plain."

The location of Argentina's Mesopotamian plain midway down the eastern coast of the continent further conforms to Solon's account:

In the centre of the whole island, there was a plain which is said to have been the fairest of all plains.[21]

As compelling as these visuals may be, dimensions are of course a far better means of verification, and Plato has much to offer in that regard. The convergence of the Paraná and Uruguay rivers marks the Mesopotamian plain's nearest approach to the sea, and Plato conveys Solon's highly restrictive measurement for the maximum distance between the plain and the sea through a collection of measurements.

Solon first specifies a distance of 5.7 miles from the plain to a small

[20] *Critias* 117e, 118a
[21] Ibid., 113c

mountain or hill from which was formed a concentrically zoned capital city:

> *Near the plain again, and also in the centre of the island at a distance of about fifty stadia [5.7 miles], there was a mountain not very high on any side Poseidon breaking the ground, enclosed the hill, making alternate zones of sea and land larger and smaller, encircling one another; there were two of land and three of water, which he turned as with a lathe, each having its circumference equidistant every way from the centre.*[22]

Solon also specifies a distance of 5.7 miles between the capital city and the sea, the length of a canal connecting the two:

> *Beginning from the sea they bored a canal of three hundred feet in width and one hundred feet in depth and fifty stadia [5.7 miles] in length, which they carried through to the outermost zone, making a passage from the sea up to this, which became a harbour, and leaving an opening sufficient to enable the largest vessels to find ingress.*[23]

This provides a combined distance of 11.4 miles with the city in the center filling the remaining span. This concentrically zoned city itself measured 27 stadia or 3.1 miles across (Fig. 49):

> *Moreover, they divided at the bridges the zones of land which parted the zones of sea, leaving room for a single trireme to pass out of one zone into another they covered over the channels so as to leave a way underneath for the ships; for the banks were raised considerably above the water. Now the largest of the zones into which a passage was cut from the sea was three stadia in breadth, and the zone of land which came next of equal breadth; but the next two zones, the one of water, the other of land, were two stadia, and the one which surrounded the central island was a stadium*

[22] *Critias* 113c

[23] Ibid., 115d

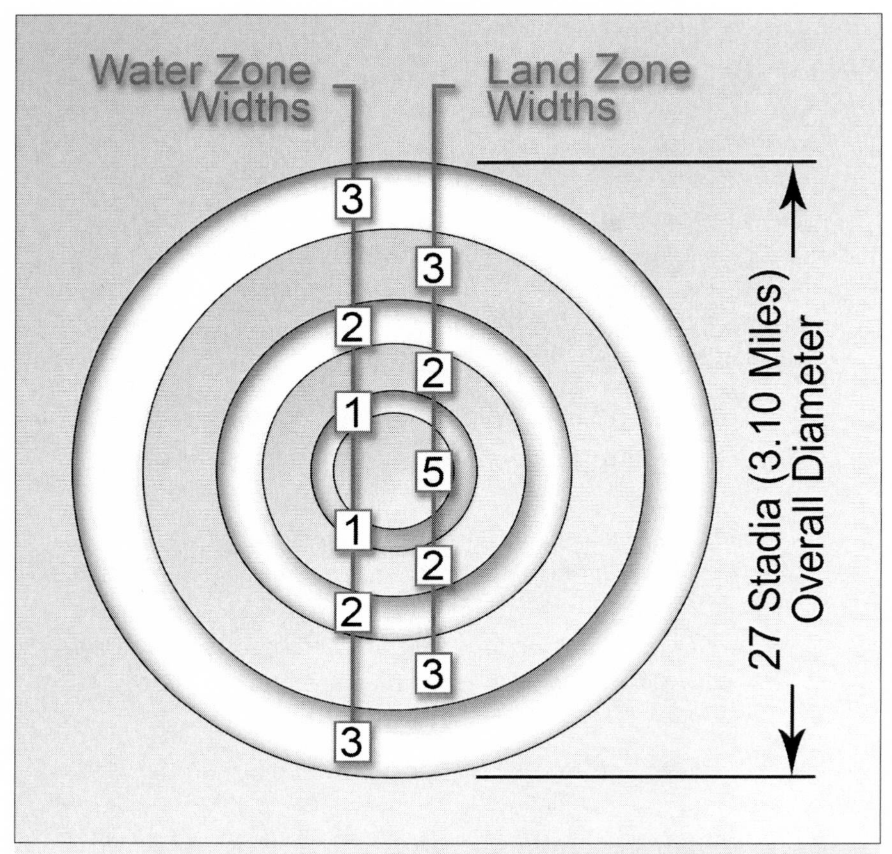

Figure 49. *The concentrically zoned city overlaid with zone widths in stadia. Zones combine to form a 27-stadia (or 3.1-mile) overall diameter. Combining this diameter with the 5.7-mile distance from the plain to the city and the 5.7-mile distance from the city to the sea establishes that the plain was located within 14.5 miles of the sea. This measurement is a useful tool in verifying any potential Atlantis plain.*

> *only in width. The island in which the palace was situated had a diameter of five stadia.*[24]

The combined measurement of 14.5 miles (5.7 miles from plain to city + 3.1 mile diameter city + 5.7 miles from city to sea) sets an extremely strict qualifier that could readily discredit this or any other posited site. As an example, consider the only other significant plain

---

[24] *Critias* 115d, e

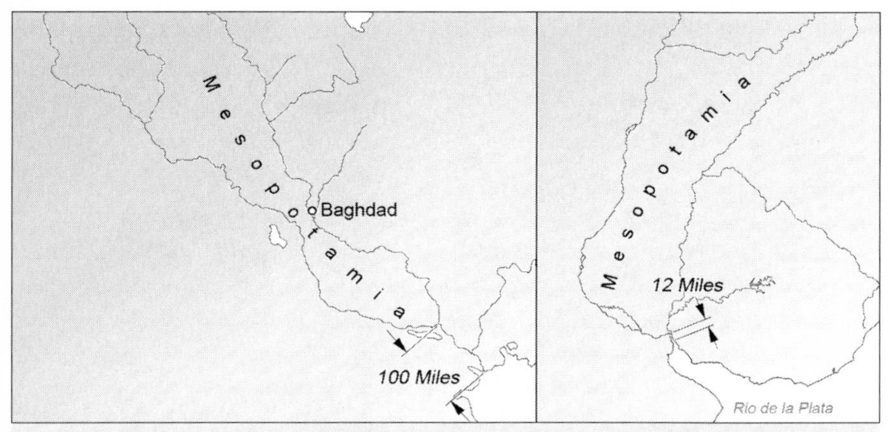

Figure 50. *The original Mesopotamian plain in Iraq (left) lies over 100 miles inland, coming nearest the sea where the two outlying rivers converge. This places it well outside Solon's 14.5-mile qualifier, easily negating it as a possible site for the Atlantis plain. Argentina's Mesopotamia (right), however, lies within 12 miles of the sea, like its namesake, being closest to the sea where its outlying rivers converge. This places South America's Mesopotamia within Solon's 14.5-mile range.*

defined by navigable waterways, Iraq's Mesopotamian plain. It lies nearest the sea where the outlying waterways of the Tigris and Euphrates converge, but the distance between plain and sea is 100 miles, meaning in this instance the city lying 5.7 miles from the sea would be located over 90 miles from the plain, or 90 miles off the mark (Fig. 50).

Adding significantly to the evidence pool, we find the convergence of the two waterways that form Argentina's Mesopotamian plain lying at a distance ranging between 8 to 12 miles from the sea, well within the account's specified 14.5-mile range.

Concession 1: Proportion

The incredible odds of a real-world plain reaching this level of conformity with Atlantis are staggering. Yet not all is perfect with this proposed site of the Atlantis plain. All site theories have had to make some concessions and this one is no different. However, I believe the few discrepancies that remain can be reasonably resolved within the context of Plato's dialogues.

The first discrepancy I will address is the Mesopotamian plain's proportion. At first glance the plain appears to be aligned accurately, with its length running parallel to the coast and its shorter width extending inland from the coast:

> *The country immediately about and surrounding the city was a level plain, itself surrounded by mountains which descended towards the sea; it was smooth and even, and of an oblong shape, extending in one direction three thousand stadia, but across the centre inland it was two thousand stadia.*[25]

However, even though the perimeter is of the correct length and the alignment true, there is a proportional disparity. Rounded to the nearest thousand, the Mesopotamian plain would be more accurately estimated at 4,000 by 1,000 stadia, whereas the Atlantis plain was specifically described as having a relatively wider 3,000 by 2,000-stadium proportion (Fig. 51).

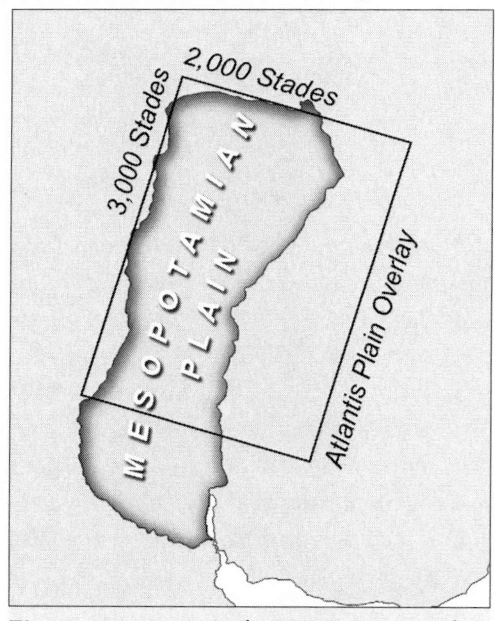

Figure 51. *Argentina's Mesopotamian plain with a scaled overlay of the Atlantis plain. While there is clear disparity in proportion, there is also evidence that the introduction of dimensions for width and length may have been assumptions added into the account by Solon.*

This discrepancy in length and width is far simpler to overcome than it would initially appear. There is a very strong possibility that the original source may have only provided an estimated measurement for the rectangular plain's perimeter and described its alignment parallel with the coast, while a later source introduced mea-

[25] *Critias* 118a

surements of width and length. If an individual felt it necessary to introduce these dimensions, given the overall measurement of the rectangle at 10,000 stadia—a length clearly rounded to the nearest thousand—he would have likely chosen width and length from the two available options of numbers, similarly rounding to the nearest thousand, that being 2,000 by 3,000 or 1,000 by 4,000. In this particular instance, he may have favored a measurement of 2,000 by 3,000.

As noted earlier, the Atlantis account was originally conveyed to Solon by an Egyptian priest. Critias states very clearly that Solon took some liberties with the original account:

> *Before proceeding further in the narrative, I ought to warn you, that you must not be surprised if you should perhaps hear Hellenic names given to foreigners. I will tell you the reason of this: Solon, who was intending to use the tale for his poem, enquired into the meaning of the names, and found that the early Egyptians in writing them down had translated them into their own language, and he recovered the meaning of the several names and when copying them out again translated them into our language. My great-grandfather, Dropides, had the original writing, which is still in my possession, and was carefully studied by me when I was a child. Therefore if you hear names such as are used in this country, you must not be surprised, for I have told how they came to be introduced.*[26]

Solon was tailoring his retelling of the Egyptian account to an Athenian audience by replacing Egyptian names with Greek ones. Yet the substitution of foreign names with Greek was not the only Hellenized aspect of the account. Solon also introduced the trireme, an ancient Greek warship, into the narrative. Solon even goes so far as to stock the ship with the standard Greek 200-man crew:

> *The docks were full of triremes and naval stores, and all things were quite ready for use.*[27]

---

[26] *Critias* 113a, b
[27] Ibid., 117d

Each of the lots in the plain had to find a leader for the men who were fit for military service, and the size of a lot was a square of ten stadia each way, and the total number of all the lots was sixty thousand. And of the inhabitants of the mountains and of the rest of the country there was also a vast multitude, which was distributed among the lots and had leaders assigned to them according to their districts and villages. The leader was required to furnish for the war the sixth portion of a war-chariot, so as to make up a total of ten thousand chariots; also two horses and riders for them, and a pair of chariot-horses without a seat, accompanied by a horseman who could fight on foot carrying a small shield, and having a charioteer who stood behind the man-at-arms to guide the two horses; also, he was bound to furnish two heavy armed soldiers, two slingers, three stone-shooters and three javelin-men, who were light-armed, and four sailors to make up the complement of twelve hundred ships. Such was the military order of the royal city.[28]

There are a few details here suggesting that Solon introduced this passage as a means of conveying the power and size of the Atlantean force in familiar Greek terms. All this relied entirely on establishing the area—width times length—of the plain; the perimeter of the plain alone would not have provided the two mathematical factors necessary to calculate the area and capacity of the plain. By assuming a 2,000 by 3,000 dimension for the plain, Solon is able to define the existence of 60,000 lots of 10 square stadia (versus 40,000 lots, should the plain have measured 4,000 by 1,000). With each lot providing 4 sailors, he then conveniently stocks 1,200 Greek triremes with standard Greek crews of 200 men. (60,000 lots x 4 sailors ÷ 1,200 triremes = 200 men per vessel.)

Of course, it is doubtful that the Atlanteans used triremes, and perhaps too much to assume that the ships they did use would be crewed identically to Greek ships. Triremes were most likely not in use at the time, and if they had been, they would have required a design much different from the Greek version. Triremes were designed with three banks of oars on each side (Fig. 52) that when properly manned and operated made the vessels extremely swift and maneuverable in battle. Greek triremes, however, were typically operated on calm seas near the

[28] *Critias* 118e-120d

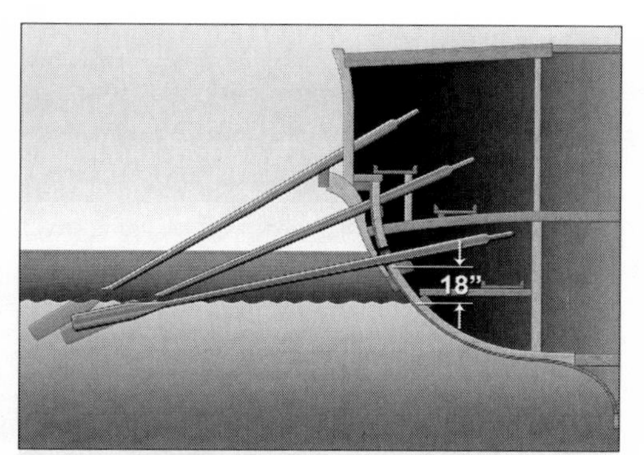

Figure 52. *Cross-section of the ancient Greek warship known as the trireme. The trireme was propelled by three banks of oars on each side of the ship, which made it both swift and agile. This design would have made it impractical for a people undertaking transatlantic voyages as the lowest bank of oars sat within 18 inches of the waterline, rendering it susceptible to sinking in rough waters.*

coast. The reason for this is that the lowest set of oar holes could lie within 18 inches of the waterline when fully crewed, making them highly susceptible to sinking should rough seas wash through the oar holes and fill the ship's hull.

Adding to the evidence that this was a personal attempt by Solon to estimate the potential size of the Atlantean force is the fact that he estimates the number of combatants based on all 60,000 lots being populated and all lots being the same large proportion. While it is remotely possible that all the lots just happened to be of the same large size and fully populated at the time of the conflict, it seems more feasible that Solon populated all the lots himself and set the larger size of the lots to effect a seemingly conservative yet imposing estimate of the Atlantean force.

From all appearances, there is ample evidence that Solon may have been compelled to insert an (inaccurate) approximation of the plain's length and width into the account in order to formulate a size for the Atlantean forces, thus enhancing his epic tale.

Concession 2: Artificial Versus Natural

The second discrepancy in the plain appears to be in the naturally formed rivers defining its perimeter, a perimeter which was said to be an artificial man-made ditch.

> *I will now describe the plain, as it was fashioned by nature and by the labours of many generations of kings through long ages. It was*

> *for the most part rectangular and oblong, and where falling out of the straight line followed the circular ditch. The depth, and width, and length of this ditch were incredible, and gave the impression that a work of such extent, in addition to so many others, could never have been artificial. Nevertheless I must say what I was told. It was excavated to the depth of a hundred feet, and its breadth was a stadium everywhere; it was carried round the whole of the plain, and was ten thousand stadia in length.*[29]

I believe Critias is on the right track when he questions whether this waterway could truly be artificial. Digging a ditch 1,150 miles in length would be an astounding feat in itself, but excavating its whole length 600 feet wide by 100 feet deep seems excessive.

First, we have to consider again whether the Atlantis plain's rectangular shape was defined by the surrounding mountains or, as I believe, by an existing waterway. If mountains defined the plain, we would have to imagine that the ditch ran along the base of the mountains on three sides and cut across the open plain in the south so that "it was carried round the whole of the plain." This appears to be what Critias pictures as he suggests that the whole of the ditch required digging. Of course, there is no such mountain-bound rectangular plain in existence, configured in this manner, at this scale.

This leaves us with option two: that the plain was originally or naturally defined by an existing waterway. As I noted earlier, there already had to be an existing natural waterway on the Atlantean plain in order to provide a constant supply of water to the navigable ditch. And it is stated that the plain "was fashioned by nature." Bury's translation reinforces the idea that the plain was from the beginning a naturally formed rectangle for the most part, but a ditch was dug to correct its shape:

> *It was originally a quadrangle, rectilinear for the most part, and elongated; and what it lacked of this shape they made right by means of a trench dug round about it.*[30]

---

[29] *Critias* 118c, d

[30] *Critias* 118c, d; Bury translation.

If the Atlantis plain were similar to Argentina's Mesopotamian plain, originally or naturally rectangular by way of bordering waterways, then it is far more likely that the actual excavation taking place to correct the rectangle where "it lacked of this shape" refers to closing the only natural opening in the rectangular waterway. This would have entailed digging a ditch roughly 70 miles in length to connect the Paraná and Uruguay rivers in the north (Fig. 53). The unrealistically herculean task of excavating a 1,150-mile channel completely around the plain can suddenly be put aside for a far more reasonably sized and extraordinarily practical endeavor.

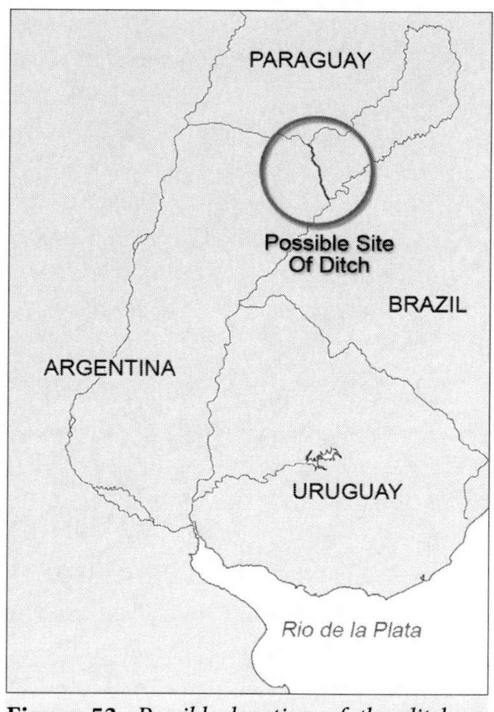

Figure 53. *Possible location of the ditch or canal that corrected the flaw in the naturally rectangular-shaped plain by closing off the plain's one opening so that water completely encircled it. "What [the plain] lacked of this shape they made right by means of a trench dug round about it."*

What makes this an exceedingly credible undertaking is that there is precedence in other ancient riverine cultures of having excavated canals of similar length for the purpose of improving irrigation and creating shorter, more efficient routes between two separate waterways. In the Middle East, workers effectively encircled the southern portion of Mesopotamia by digging a ditch roughly 40 miles in length between the Euphrates and Tigris Rivers, which became known as the *Nahar Malcha*, Royal Canal. Herodotus, a Greek historian from the fifth century BC, claimed it was of adequate width and depth to allow navigation by merchant ships.

Pliny the Elder writes about similar attempts to connect the Nile to the Red Sea as far back as the nineteenth century BCE:

> *Next comes the Tyro tribe and, on the Red Sea, the harbour of the Daneoi, from which Sesostris, king of Egypt, intended to carry a ship-canal to where the Nile flows into what is known as the Delta; this is a distance of over 60 miles. Later the Persian king Darius had the same idea, and yet again Ptolemy II, who made a trench 100 feet wide, 30 feet deep and about 35 miles long, as far as the Bitter Lakes.*[31]

There is another clue that this was indeed the original idea conveyed: the reference to "circle" and "round" in the two translations. Jowett appears to be struggling with his assumption that the excavators were straightening the lines of the entire plain and renders a somewhat confusing translation: "Where falling out of the straight line followed the circular ditch." Bury, who appears to have a better grasp on the intent of the passage, does not assume that the shape referred to is a line: "What it lacked of this shape they made right by means of a trench dug round about it."

The Greek word being translated is *kukló*, which means to circle, round about, or encircle. This allows for the following more refined translation:

> *Where the plain lacked its "rectangular" shape, they corrected it by digging a trench to fully encircle it. - Author's translation*

This ditch would not only have been highly useful for irrigation purposes; it also would have allowed full navigation around the plain, providing a more efficient alternative to overland transport in the northern region.

I contend that if Critias had realized that Solon or the Egyptians were describing a plain shaped by a natural outlying waterway, he would have readily interpreted the ditch in this same manner. Vagueness in the wording likely led to his erroneous vision of a wildly unbelievable excavation project, which he himself could scarcely fathom.

---

[31] Pliny, *Natural History* 6.33

Concession 3: Island Versus Extended Land

The third seeming discrepancy regards the whole of Atlantis. Some may have noticed my use of the word "continent" in relation to Atlantis. The reason for this is obvious: Atlantis earns that title by being an island the size of two continents:

> *Atlantis, which, as was saying, was an island greater in extent than Libya and Asia.*[32]

Yet, as you may have also noticed in this quote, Atlantis is clearly referred to as an island, making it an island continent. So how could South America have been Atlantis when it does not classify as an island, at least by our modern definition of the term? This discrepancy becomes easy to resolve when we consider how the Greeks would have defined it. There exists proof that ancient Greece would have referred to South America as an island.

Figure 54. *Peloponnese, which contains a root form of the Greek* nesos *and translates as "Island of Pelops," confirms that the Greek term* nesos *denotes a landmass surrounded by water on all sides, even if tethered to a larger landmass by an isthmus.*

In this instance, the Greek word used for island is *nesos*, and in virtually every instance it does indeed denote the same familiar meaning: a body of land completely surrounded by water. However, there is one very prominent instance where the term is applied to a slightly different landform. Peloponnese, also referred

[32] *Critias* 108e

to as Peloponnesus, is a large peninsula extending from the southern end of Greece (Fig. 54). While today we classify Peloponnese as a peninsula, a small landmass extending out from a larger landmass, the Greeks were apparently comfortable with referring to it as an island. The name Peloponnese, which contains a root form of *nesos*, translates as "Island of Pelops." This term acknowledges the peninsula's conquest by the mythical Greek hero Pelops, while also acknowledging that ancient Greece did indeed refer to at least one peninsula as a *nesos*, or island.

Figure 55. *The Atlantis 'nesos'. Like Peloponnesos, South America is tethered to a continent by a narrow isthmus, in this case the Isthmus of Panama. Also like Plato's Atlantis, a path of islands—the Caribbean Islands—lead to a continent on the opposite end, North America. The Caribbean Islands would have proved a very efficient route to North America for a maritime people dwelling in the vicinity of Rio de la Plata.*

This instance suggests that *nesos* may have had a slightly broader meaning describing any landmass surrounded by water on all sides. Unlike typical peninsulas, which extend out from a larger landmass with water washing up on three shores, Peloponnesos is bound by water on all sides but tethered by a narrow strip to the Greek mainland.

South America shares this same geographic structure. The continent is almost entirely encircled by water, but like Peloponnesos, it is tethered to a larger landmass by a long narrow isthmus—Panama in this instance. The term *nesos* may therefore be applied to South America in the same sense.

Another very good possibility is that someone, perhaps Solon, may have incorrectly inferred that Atlantis was an island. The account describes two continents; Atlantis and another large continent, which was accessible via a path of islands. This would be a very good description of South America with the Caribbean Islands forming a very distinct path to the North American continent on the opposite end (Fig. 55):

> *The island was larger than Libya and Asia put together, and was the way to other islands, and from these you might pass to the whole of the opposite continent.*[33]

While the original story may have simply pointed out the islands' significance as an important, unique route between continents, it would have been very easy for someone to misconstrue the inclusion of this sea route between continents as the only route, hence Atlantis interpreted as a separate island continent. In reality, the slow, tedious overland route up the Isthmus of Panama was likely omitted only because it held little relevance for a maritime civilization.

This and the previous discrepancies appear to have occurred as the account was passed down over generations. Critias refers to these multiple retellings and his difficulty in accurately recounting and reconciling the story told him in his youth. It has all the makings of an ancient game of Chinese whispers.

> *Then listen, Socrates, to a tale which, though strange, is certainly true, having been attested by Solon, who was the wisest of the seven sages. He was a relative and a dear friend of my great-grandfather, Dropides, as he himself says in many passages of his poems; and he told the story to Critias, my grandfather, who remembered and repeated it to us.*[34]

> *I did not like to speak at the moment. For a long time had elapsed, and I had forgotten too much.* [35]

> *If I (Critias) can recollect and recite enough of what was said by the (Egyptian) priests and brought hither by Solon.*[36]

> *This I infer because Solon said…*[37]

---

[33] *Timaeus* 24e
[34] Ibid., 20d, e
[35] Ibid., 26a
[36] *Critias* 108d
[37] Ibid., 110a

Perhaps the most intriguing example of how the tale was modified after it was told to Solon and before Plato put it in writing pertains to the most conspicuous of discrepancies: Unlike Atlantis, South America still lies above the sea.

CHAPTER 7

THE CONTINENT THAT SOLON SUNK

Atlantis, which, as was saying, was an island greater in extent than Libya and Asia, and when afterwards sunk by an earthquake, became an impassable barrier of mud to voyagers sailing from hence to any part of the ocean.[38]

In Plato's account, Solon informs us that Atlantis—an island continent greater in size than Asia and Libya combined—sank beneath the sea in the span of a day. However, a key discovery detailed in this section exposes the circumstances that influenced this clearly false assumption, while also demonstrating the legitimacy of an unsubmerged South America as the possible site of Atlantis.

It all came to light when I focused my attention on Solon's comparison of the Mediterranean Sea with another body of water known as the "true ocean" and the sudden introduction of a new and mysterious "boundless continent." Following is the relevant passage from Timaeus, with my interpretation of the described landforms (reference Figure 55) inserted in brackets:

The island [Atlantis/South America] *was larger than Libya*

[38] *Critias* 108e

> *and Asia put together, and was the way to other islands* [the Caribbean Islands], *and from these you might pass to the whole of the opposite continent* [North America] *which surrounded the true ocean; this sea which is within the Straits of Heracles* [the Mediterranean] *is only a harbour, having a narrow entrance, but that other is a real sea, and the surrounding land may be most truly called a boundless continent. In this island of Atlantis there was a great and wonderful empire which had rule over the whole island* [Atlantis/South America] *and several others* [the Caribbean Islands], *and over parts of the continent* [North America].[39]

If you compare other theories that attempt to reconcile their Atlantis theories with a boundless continent surrounding the "true ocean," you will find they abandon historical context and reconcile the "true ocean" and a surrounding continent to the world as we know it today. Rand and Rose Flem-Ath attempt to validate their popular hypothesis that Atlantis is Antarctica by suggesting that Solon was describing the oceans surrounding Antarctica with all other continents composing the opposite continent. Jim Allen who, like me, maintains that South America is Atlantis—albeit with its capital city located high in the Andes Mountains—supposes that the "true ocean" was the Pacific while Eurasia represents the opposite continent that could be reached via the many islands of the Pacific. Unfortunately neither of these interpretations define an ocean surrounded by one large contiguous or boundless continent.

As I said, these theorists arrived at their conclusions through a natural and instinctive process of reasoning and reconciled an ancient Atlantis account to their own modern worldview. So why would we not suspect Solon of having done the same, rationalizing an ancient Egyptian account within his own worldview?

Having attained a familiarity with ancient Greek maps, I came to realize that Solon was providing a slightly skewed description of the ancient Greek worldview. Greek maps comprised two large bodies of water and one large landmass (Fig. 56). The two main bodies of water were the Mediterranean Sea (*Mare Magnum*), which is encircled by the

---

[39] *Timaeus* 24e-25a

Figure 56. *Reconstruction of Hecataeus' world map, which depicts the ancient Greek worldview of two major bodies of water, the Mediterranean Sea and* Oceanus, *separated only by the Pillars of Hercules, or the Strait of Gibraltar as we know it today.*

three combined continents of Europe, Libya, and Asia, and the much larger "true ocean," Oceanus, surrounding the three continents. The only link between these two seas is the "narrow entrance" into the Mediterranean Sea known then as the Pillars of Hercules (and today as the Strait of Gibraltar). This ancient Greek worldview complies perfectly with Solon's description of the two linked large bodies of water.

The skewed portion of Solon's description was the sudden introduction of a continent completely surrounding Oceanus. It was a concept I was not too familiar with, but I had happened upon a reconstruction of a Greek map that depicted this outer continent. The reconstruction portrays the world according to sixth-century Greek philosopher Anaximander. The world takes the form of a large cylinder that is a third tall as it is wide. Covering one of the cylinders flat surfaces is a large body of water representing Oceanus and like other maps of the day, Europe, Libya, and Asia are combined into a circular landmass centered in its midst. The sides of Anaximander's cylinder are raised all along the perimeter creating a wispy mountainous barrier and an outer shore fully encircling Oceanus. This portrayal matches Solon's description of the world perfectly.

Most of Anaximander's work has been lost to us. All that remains are references and descriptions provided by classical historians. Reconstructions of the world's structure like the one described are based on interpretation of what little information exists. There is no certainty that Anaximander incorporated an outer continent on his map, but it is a concept that would not have been completely foreign to him. Anaximander was familiar with Homer, a Greek legend, whose epic tales had become an essential part of Greek education and culture. Homer makes several references to the existence of a land lying on the other side of Oceanus. He believed Hades, the land of the dead, could be reached via this distant land.

> *But when in thy ship thou hast now crossed the stream of Oceanus, where is a level shore and the groves of Persephone—tall poplars, and willows that shed their fruit—there do thou beach thy ship by the deep eddying Oceanus, but go thyself to the dank house of Hades.*[40]

Solon is clearly building on Homer's concept of a land beyond Oceanus by stating that this far-off land extended completely around Oceanus. Since there are no continents surrounding an ocean in the real world and the only historical descriptions of such a landform lie in a worldview existing in Solon's time and an account of Atlantis also from Solon's time, simple logic dictates that these two boundless continents surrounding a large "true ocean" were one and the same.

---

[40] Homer, *Odyssey* 10.508

It is a virtual certainty that Solon was reconciling the Egyptian account of two large continents in the waters beyond the Mediterranean Sea to his own worldview. The problem with this limited view was that the Greeks believed Oceanus was an enormous river forming a continuous stream around the known world with only a few small islands scattered about. According to their reasoning, there were no continental landmasses breaching its surface and impeding its flow.

Solon could associate the Egyptians' inclusion of an opposite continent (North America) to the outer ring of shoreline surrounding Oceanus as this would have been the only other continent-sized landmass known to be raised above the sea besides Europe, Libya, and Asia. This mistaken association would explain the altered description of North America. Solon's worldview maintained a specific order with the circular form of the inhabited earth encircled by Oceanus which was in turn encircled by a ring of land. There existed no other landform within this worldview that could be reconciled to the large continent-sized landform, Atlantis.

According to Solon's account, the Egyptian priest described not only the continent as an island but the capital city as an island as well. The priest related that "in a single day and night" the island had "disappeared in the depths of the sea." I believe that the island that actually sank was the capital city. Solon, uncertain how to reconcile the existence of a massive new continental landmass within his worldview, would eventually find resolution by interpreting the Egyptian account of a sinking island, which was intended to be limited to the small island capital, as the sinking of the entire island continent of Atlantis. By interpreting the Egyptian account in this way, Solon would be able to maintain the Greek worldview with its limited landforms and retain an unimpeded "ocean stream" that flowed over a submerged Atlantis, as demonstrated in Figure 57.

Fully supporting this possibility are similar adjustments made in modern times as in the case of the small island of Santorini in the Mediterranean. Modern-day researchers, similar to Solon, have applied similar grand-scale alterations to the size and location of Atlantis. They have raised it up from the depths of the Atlantic and relocated it within the Mediterranean, reducing its size from that of a continent to a minuscule island, all in an effort to reconcile the possible reality of an ancient foreign account within the limits of their current worldview. Unable to accept either the sinking of a continent or the existence of a highly advanced maritime culture existing beyond the Mediterranean, they have reinterpreted the account to suit what they believe are established, inalterable historical principles.

A curious side note: Solon's reconciling Atlantis to Homer's or Anaximander's worldview does add a layer of authenticity to a tale many consider fiction. Plato described two continents beyond the known world in terms that most of his contemporaries would not have understood. He described them as they would be arranged onto a circular disk, but he and his contemporaries held to a more modern worldview of a spherical Earth. It is a cryptic description that would have been natural in Solon's day but lost on Plato's contemporaries, just as it has been lost to most of the world until now.

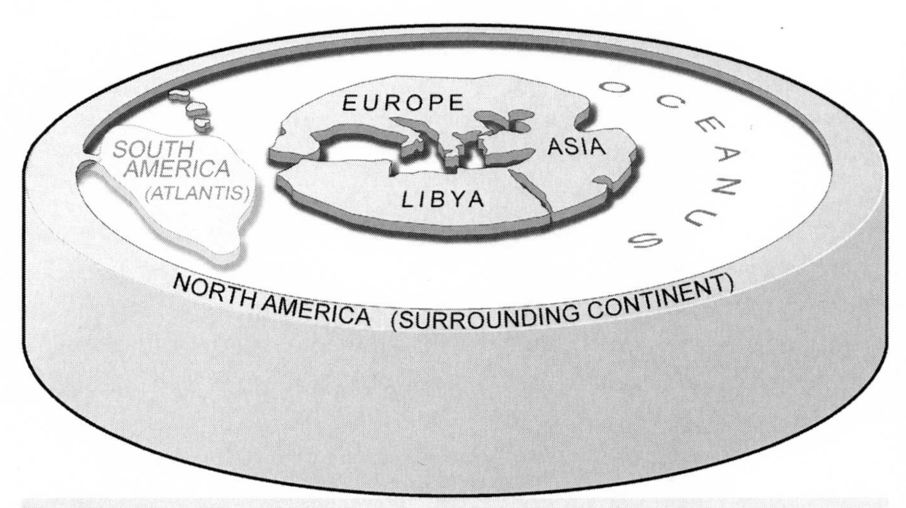

Figure 57. *Author's reconstruction of Solon's worldview. A worldview that Solon apparently adapted to an Egyptian tale of two large continents lying beyond the Mediterranean. He determined that the larger continent encircled Oceanus while the other, Atlantis, rested beneath it.*

A City Sinks

Lending support to the view that the sinking applied to the capital city and not the entire continent is the high likelihood that the city was built on a delta. Suddenly the destruction becomes far clearer.

> *There occurred violent earthquakes and floods; in a single day and night of misfortune all your warlike men in a body sank into the earth, and the island of Atlantis in like manner disappeared in the depths of the sea. For which reason the sea in those parts is impassable and impenetrable, there is a shoal of mud in the way; and this was caused by the subsidence of the island.*[41]

Were a series of "violent earthquakes" to have occurred, as Solon relates, the impact on a river delta would have been catastrophic with results matching very closely Solon's description. During an earthquake, a phenomenon known as liquefaction can occur, destabilizing the ground and transforming the soil into a liquid consistency.

The silt and sand composition within a delta would have rendered the capital city and surrounding area highly susceptible to liquefaction and may explain why the "warlike men" were said to have sunk into the earth when the "violent earthquake" occurred, as opposed to having sunk into the sea. The agitated soil would have also transformed the delta's many distributaries and the channel said to lead toward the city into mudflows—thick rivers of mud that would have left "an impassable barrier of mud to voyagers."

One only has to look as far as the slowly sinking delta city of New Orleans and imagine the impact a strong quake would have on such a site. The resulting cataclysm would likely replicate the destruction recounted in Plato's dialogues, with a massive city sinking into the earth and disappearing beneath the sea in a single day and night.

The determination of Atlantis' capital city location in a delta begins where Solon tells us that the surrounding waterway "winding round the plain and meeting at the city, was there let off into the sea." We know

---

[41] *Timaeus* 25c, d

that the waters did not meet precisely at the city, because Solon held that the city was located 50 stadia (5.7 miles) from the plain and locating the city immediately at the point of convergence would place it squarely on the plain's border. Perhaps acknowledging and reconciling the passage within this context, R.G. Bury offers the following alternate translation:

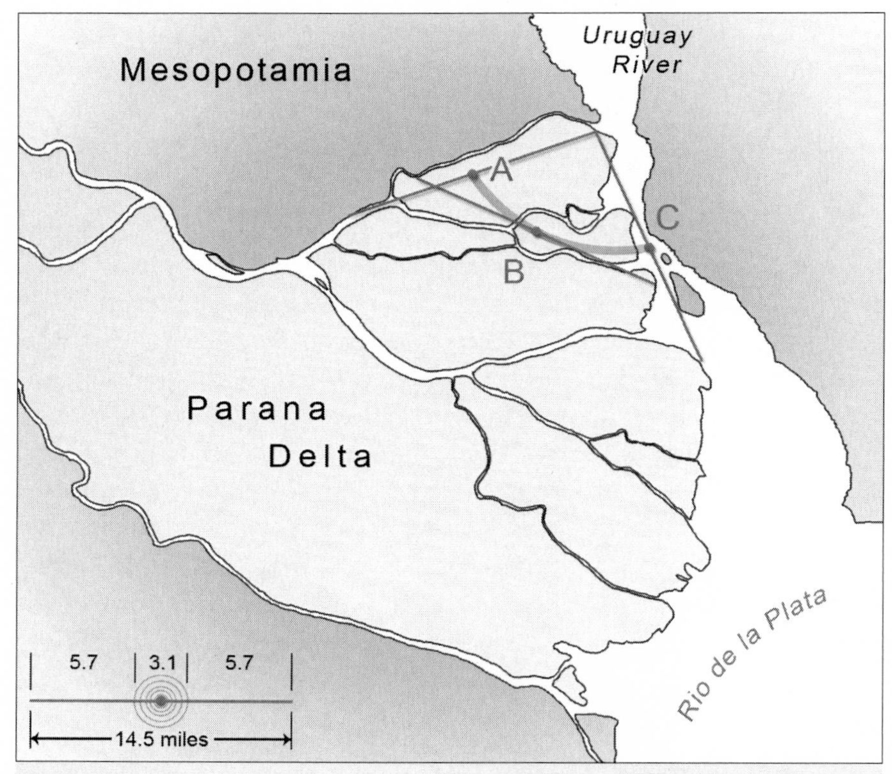

Figure 58. *Lines A, B, and C represent 14.5-mile channels extending between the plain and the sea with a center point at 7.25 miles representing the center point of the island city. Lines A and C represent extreme channel placements where the island city would lie closest to the Rio Gutierrez and the confluence of the Rio Gutierrez and Uruguay rivers. Line B aligns the channel so that the center point is equidistant from the Mesopotamian plain and the sea. Multiple other channel placements would find the city's center located between these three points along or near arc ABC. Equidistant point B mysteriously sits atop a circular landform, while line B itself lies near channels of similar overall length leading from the plain to the sea. Could this be the site of Atlantis' island city?*

It received the streams which came down from the mountains and after circling round the plain, and coming towards the city on this side and on that, it discharged them thereabouts into the sea.[42]

We established that the city was the same distance from the plain as it was from the sea. This would center the city 7.25 miles—(5.7 miles from plain to city + the city's 3.1-mile diameter + 5.7 miles from city to sea) ÷ 2—between the plain and the sea. In Figure 58, an array of lines 14.5 miles in length with midpoints representing the central city extend between various points along the plain and the sea allowing for multiple locations of the island city along an arc designated ABC. The most extreme placements of this template locate the central city along Rio Gutierrez, a more northern offshoot of the Paraná River, at point A, and atop the confluence of the Rio Gutierrez and Uruguay Rivers at point C. Extending the channel between the Rio Gutierrez along the plain and the confluence so that the center point is equidistant from both waterways, we find center point B lying atop an intriguing and unique formation.

This center point sits atop a very curious circular landform in the Paraná Delta. A small distributary channel breaking off from Rio Gutierrez flows eastward alongside line B before splitting into two waterways, creating a river bifurcation. The split waterway flows almost entirely around the circular formation but veers away suddenly toward the sea just before completely enclosing the circular landform. The landform consequently appears as a three-quarter circle with a triangular extension leading off to the east.

There are several levels of intrigue here:

1. It is the only landform in the Paraná Delta approaching a circular form.
2. It happens to fall very closely within the limited range of possible locations afforded the city of Atlantis per Plato's narrative, and
3. This is perhaps the most amazing aspect to consider. The circular portion of the landform conforms very closely to Solon's dimensions for the circular city (Fig. 59).

---

[42] *Critias* 118d; translation by R.G. Bury

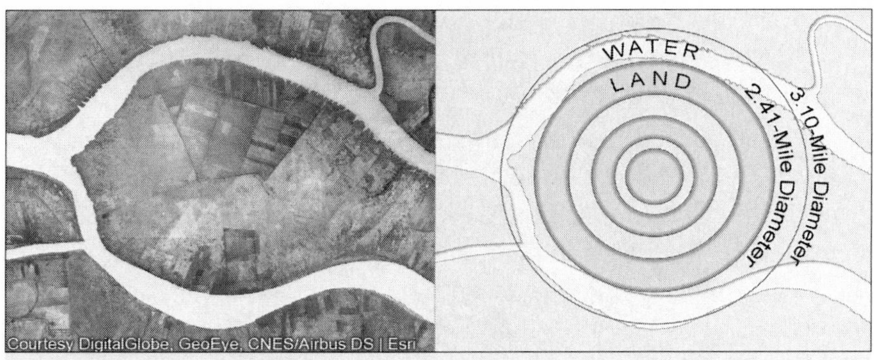

Figure 59. *Satellite image of the circular landform in the Paraná Delta (left). Same image (right) with overlain rings conforming to Atlantis' concentric rings of land and water scaled to the image. While the circular landform does not appear to have any visible demarcations suggesting the existence of inner zones or rings, the outer zone of water conforms fairly closely to the specified 2.41-mile inner and 3.1-mile outer diameters. If this is the site of the city, settling of sediment into the sunken city's outer channel may have deformed the delta's surface just enough, allowing water flowing through the delta to partially retrace the outer channel's circular shape.*

Solon's combined measurements for the multi-ringed city establishes the overall diameter at 27 stadia, or 3.1 miles. Solon claimed, however, that this first outermost ring was composed of a 3-stadia wide channel of water. This gave the next concentric ring of the city, which was land, an overall diameter of 21 stadia, or a 2.41-mile diameter for the largest ring of land. Both diameters match up very closely with the circular portion of this Paraná Delta landform. Could this circular formation possibly be the site of the legendary city?

Water does not choose a course entirely at random; gravity and terrain are key in determining its path. If the city had found itself eventually buried beneath layers of silt, it stands to reason that sediment would likely have settled deeper into the ringed channels, whose beds sat lower than the rings of land. This may have in turn formed shallow troughs on the delta's surface. Water flowing back into the area could encounter the surface deformations and be redirected around the sunken city, but it is also reasonable to expect that it would not necessarily follow the shallow deformation entirely and perfectly retrace the original circular channel. The waterway's sudden break from a full circle in the east may suggest the momentum of the water flowing through the trough was too great, allowing the water to break out of the outer ring's contour. This would result in the current three-quarter circular formation with land extending off to the east.

There are some very slight discrepancies in this theory. Still, the central circular feature presents a thought-provoking conundrum. I have studied other deltas throughout the world, many of which contain hundreds of islands and river bifurcations. None possess landforms nearly as round, and none approach this size. Given all the evidence presented thus far with the Mesopotamian plain's proper alignment, size, positioning, and distance from the sea, this landform is definitely a site of significant interest in our search for Atlantis.

CHAPTER 8

THE LONG WALLS OF ATLANTIS

Although this circular feature is located almost precisely where the account suggests the city would lie in relation to plain and sea, it is important to keep in mind that our earth is dynamic and ever-changing, leaving other factors still at play. For example, changes in the region over thousands of years may have altered the waterways. In other words, the delineation between the plain and the delta may be substantially different from what it was in the past, which could place the original site for the city miles from the site currently proposed.

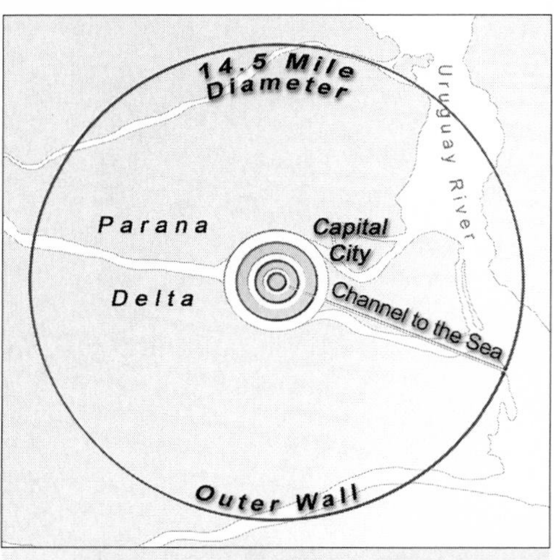

Figure 60. *A bad fit? The capital city of Atlantis laid out on the Paraná Delta with its proposed outermost wall measuring over 45 miles in circumference. The placement of the wall, based on the location of the circular landform, sees it straddling the Uruguay River in two places.*

One of the questions that arises about this location is that there is limited space for the outer wall that most believe completely encircled the city. This wall was said to have an overall diameter of 14.5 miles (Fig. 60).

However, as I will demonstrate, this vast outer wall was never actually described in ancient accounts of the city. It is simply a common misconception that originated as a result of the translators overlooking contextual continuity within Plato's account. Regardless of whether or not Atlantis existed, Critias, who recounted the tale, would have had a clear vision of what he was attempting to portray. The following analysis will dive deep into the text and help find that original vision.

Here is the passage in dispute:

> *And after crossing the three outer harbors, one found a wall which began at the sea and ran round in a circle, at a uniform distance of fifty stadia from the largest circle and harbor, and its ends converged at the seaward mouth of the channel.*[43]

It would appear from this translation that Critias was indeed describing a wall that fully encircled the circular capital city at a uniform distance, paralleling its outermost ring at a distance of 50 stadia (5.7 miles), with its ends converging at the sea. As the city had a diameter of 3.1 miles, this would put the wall at 14.5 miles in diameter and over 45 miles in circumference.

It is certainly an impressive structure, yet one based on an imagined fourth wall, where the original account is clearly defining the city's outermost third wall and a walled channel extending off it. Let us consider the first portion of the passage:

> *And after crossing the three outer harbors, one found a wall which began at the sea.*[44]

It is clear that the translator is suggesting that "after crossing the three outer harbors, one found a wall" lying at the sea 50 stadia from the city, but contextually this is wholly incorrect. After crossing the three outer harbors, one actually came to the mouth of the 50-stadium channel within the outer harbor, not at the sea. More importantly, one

---

[43] *Critias* 117d, e; translation by R.G. Bury

[44] Ibid., 117e

did indeed come to a wall—a wall lining "the largest circle and harbor," one of three walls existing in the multi-ringed city. This conforms to a previous passage where Critias relates that the outer harbor or outermost circle was lined with a brass-covered wall:

> *And they covered with brass, as though with plaster, all the circumference of the wall which surrounded the outer-most circle.*[45]

Now many, including Bury, have incorrectly assumed that this brass wall surrounded the outermost ring of land. But as the account does not specifically state whether this circle was associated with land or water, we must adhere to contextual consistency and review other references to the phrase "outermost circle" to make the proper determination. Critias' only other reference to the "outermost circle" is clearly regarding the outermost circle of water or harbor and occurs just a few sentences earlier. It is highly doubtful that Critias would assign two different meanings to a unique phrase addressed within a single continuous thought:

> *For, beginning at the sea, they bored a channel right through to the outermost circle, which was three plethra [303 feet] in breadth, one hundred feet in depth, and fifty stades [5.7 miles] in length; and thus they made the entrance to it from the sea like that to a harbor by opening out a mouth large enough for the greatest ships to sail through.*[46]

It is clear here that they bored a channel through to the outermost circle of water, or "made the entrance to" the ring of water so their largest ships could access it. Shortly thereafter, the account continues:

> *And they covered with brass, as though with plaster, all the circumference of the wall which surrounded the outermost circle; and that of the inner one they coated with tin; and that which*

[45] *Critias* 116b; Bury
[46] Ibid., 115d; Bury

encompassed the acropolis itself with orichalcum which sparkled like fire.[47]

This establishes without a doubt that there was a brass-coated wall lining the outermost circle of water. But let us also take a moment to look at the positioning of the remaining two walls. Critias specifies the location of this outermost brass wall and the location of the orichalcum wall he places on the small central island or *acropolis*, but he refers to the tin wall as merely "the inner one." Benjamin Jowett provides a translation that gives a slightly more specific location for the tin wall:

The entire circuit of the wall, which went round the outermost zone, they covered with a coating of brass, and the circuit of the next wall they coated with tin, and the third, which encompassed the citadel, flashed with the red light of orichalcum.[48]

This places the tin wall in the middle of a sequence that begins with the brass outer wall and ends at the central orichalcum wall. If the brass wall had surrounded the outermost ring of land, as many have wrongly assumed, and the orichalcum wall surrounded the central island, this would establish a pattern of wall-bound zones of land. Therefore, "next" in the sequence could most definitely be discerned as the outer edge of the smaller ringed island.

However, since the brass wall actually surrounded the outermost ring of water and the orichalcum wall surrounded the central island, the sequence is not limited to the placement of walls around islands, but rather all delineations between rings of water and land become part of the sequence. Thus, "next" in the series after the brass wall would be the next delineation between land and water, establishing that the largest ring of land was bound by the tin-clad wall. The only entrance through this tin wall is where the short narrow channel restricted passage to a single trireme, demonstrating that the three islands were secure and exclusively intended for the military and royalty, while the outer harbor and channel could be fully accessed by civilian merchant ships.

The ancient Carthage harbor incorporated a similar scheme. Both

---

[47] *Critias* 116b, c; Bury

[48] *Critias* 116b, c; Benjamin Jowett

military and merchant ships could enter a walled rectangular manmade harbor designed for civilian trade, but beyond it, access was limited by a narrow passage leading into the inner military harbor. This inner harbor was a circular channel fortified by two walls. It is a design that is well suited to a civilization with a strong naval presence and thriving sea trade as was said of Atlantis.

Figure 61 represents the most likely positioning of the three walls

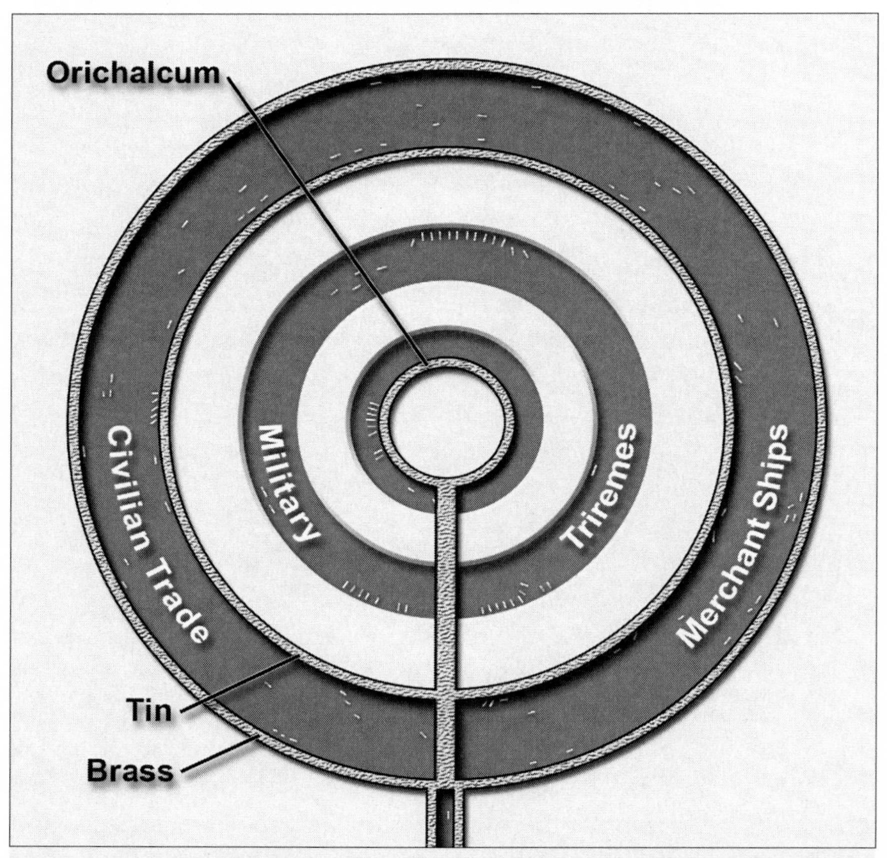

Figure 61. *This image demonstrates the most plausible placement of the three walls directly in and around the city complex. The wall of brass surrounded the "outermost circle" of water, or harbor. The tin wall followed next lining the outermost circle of land, followed by the wall of orichalcum which surrounded the citadel, the central island.*

of Atlantis' capital city. Based on this corrected layout, here again we find that after crossing the three harbors, we indeed come to the brass wall specified by Critias. If the wall being spoken of had been a fourth located at the sea 5.7 miles from the third outer harbor, a fourth wall that never came closer than 5.7 miles from the multi-ringed city and its outermost harbor, it would have only come in contact with a small portion of the channel at its seaward mouth. This misinterpretation, which sees only the ends of this far-removed wall converging with the channel at the sea, creates confusion within a portion of the passage that details interaction between those living on this wall and activity within the channel and outer harbor:

> *The whole of this wall had numerous houses built on to it, set close together; while the sea-way and the largest harbor were filled with ships and merchants coming from all quarters, which by reason of their multitude caused clamor and tumult of every description and an unceasing din night and day.*[49]

Here we have a description of the constant interaction between the inhabitants of the city dwelling on the wall and the merchant ships that filled both "the sea-way and the largest harbor." The problem is that the only such interaction with the alleged fourth wall would be limited to a very small area at the mouth of the channel near the sea where the ends of the wall converged. Therefore, there would actually be very little interaction between those dwelling on the wall and the ships entering the channel. Only a few feet of this 45-mile wall would be able to interact with the channel. Even more perplexing is why this passage would suggest interaction with the ships filling the largest harbor at the city. The ships in the harbor would have been barely visible, if at all, from any point along this vast wall. How could these distant ships motivate those dwelling on a wall 5.7 miles away to engage in loud and tumultuous behavior night and day?

If the wall being described were indeed the one encircling the outer harbor however, one can easily imagine nonstop day and night activity

[49] *Critias* 117e; Bury

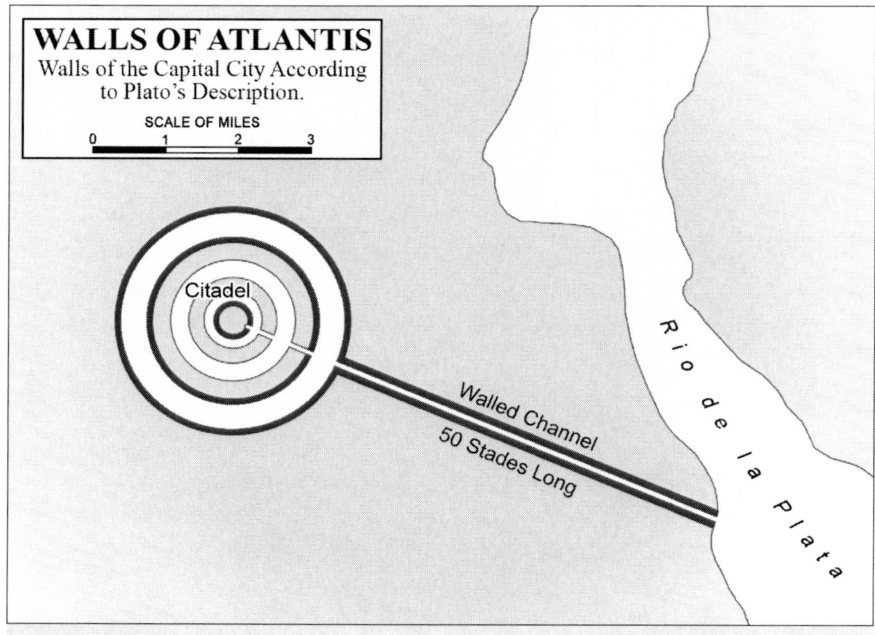

Figure 62. *Configuration of the three walls of Atlantis conforming to Critias' linking heightened day and night interactivity between inhabitants on the wall and merchant ships in both the outermost harbor and within the 5.7-mile channel. It also clarifies Critias' original vision, "After crossing the three outer harbors, one found a wall which originated at the sea a distance of fifty stades from the largest circle and harbor; It ran round everywhere with its ends converging at the seaward mouth of the channel."*

where those dwelling on the wall would be actively involved in trade with merchant ships in the harbor.

But what of the comment that the wall existed a "distance of fifty stades from the largest circle and harbor?" If, as we have established, Critias is referring to the wall surrounding the outer harbor, it becomes clear that he is describing the full extent of this same wall. He is explaining that it extends out beyond the outer harbor the length of the canal to the sea—or as he plainly states, "one found a wall which began at the sea," not a wall located *at* the sea (Fig. 62). This leads to my interpretation of the passage, which proves contextually more consistent:

And after crossing the three outer harbors, one found a wall which originated at the sea a distance of fifty stades from the largest circle and harbor; it ran round everywhere with its ends converging at the seaward mouth of the channel.

The whole of this wall had numerous houses built on to it, set close together; while the sea-way and the largest harbor were filled with ships and merchants coming from all quarters, which by reason of their multitude caused clamor and tumult of every description and an unceasing din night and day.[50]

With this new interpretation, all elements mentioned are suddenly in play and visible from this one single location at the harbor's entrance to the channel. Crossing the outer harbor and sitting in front of the entrance to the channel, Critias' audience could simultaneously perceive being fully surrounded in the harbor by a brass-covered wall while also envisioning this great structure extending far down each side of the 5.7-mile channel. From this same vantage point, we are also able to envision the many merchant ships moving about both the harbor and the channel—the harbor, and perhaps the channel to some extent, lined with docks accessible from the wall. We can envision the great amount of excitement and activity the ships' presence would generate for the multitude who dwelt on this vast wall, "clamor and tumult of every description and an unceasing din night and day."

The ancient Carthage harbor incorporated a similar scheme. Commercial and military ships could enter a rectangular manmade harbor designed for civilian trade, but beyond it access was limited and controlled leading into the inner circular military harbor, which was fortified by walls. It is a design that is well suited to a civilization with a strong navy and thriving sea trade.

Adding substantial credibility and practicality to this proposed layout is the existence of the similarly fashioned Long Walls of Athens (Fig. 63). Like the walls of Atlantis, the Long Walls of Athens encircled the city and extended outward, forming a long narrow corridor to the sea, the only difference being that Athens' passage was of land while Atlantis' passage was of water. Similar long wall constructions were established throughout Greece as a means of securing access to the sea.

---

[50] *Critias* 117d, Author's translation

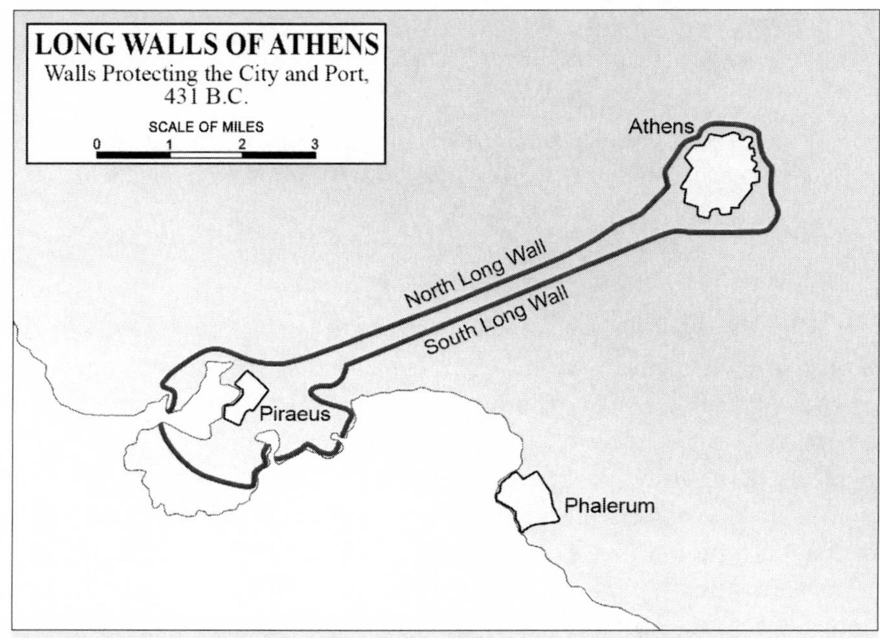

Figure 63. *The Long Walls of Athens as they existed at the time of the Peloponnesian War. Similar to the walls of Atlantis, they provided a secure narrow corridor through which the city was able to maintain access to the sea.*

In the case of the Long Walls at Athens, the walls secured a 40-stadium (4.5-mile) passage to the port city of Piraeus, providing safe transport of supplies to Athens in times of a land siege.

This shared attribute between Atlantis and Athens also introduces an interesting chicken-versus-the-egg conundrum. If the Atlantis saga is true, could Solon's description of its city walls have influenced the building of the Greek long walls a century later? Alternatively, if either Plato or Critias invented a fictitious Atlantis, did they base the design of its capital city on the existing Long Walls of Athens?

CHAPTER 9

HISTORICAL ATLANTIS

There are certainly reasons to doubt the Atlantis tale in its entirety, but there are no documents besides Plato's dialogues preceding the sixteenth century that approach this level of accuracy in describing the Americas. It appears to be the earliest recorded account recognizing the existence of not one but two continents in the Atlantic independent of the ancient known world—Europe, Asia, and Africa—and throws in an accurate description of the Caribbean islands for good measure.

Whether Atlantis is real or imagined, there should be no doubt that had it truly existed at the purported scale detailed in Plato's dialogues, then South America is indeed the site of Atlantis. Limited to the confines of our finite Earth, there is, without exception, no other site proposed or yet to be proposed that could ever conform as closely to the combination of grand scale geographic features and restrictive parameters provided in Plato's account. Whether by chance or by genuine ancient source, Plato provides us the first faithful description of the Americas nearly 2,000 years before the two continents were charted by Europeans.

If the tale is true, perhaps it is no coincidence that an Egyptian priest from the sixth or seventh century BCE should be the one to inform Solon of a civilization inhabiting the South American continent. In 1992, German toxicologist Svetla Balabanova shocked the world

Plato's America

1. North and South America—Plato acknowledges the existence of not one, but two large continents in the Atlantic.

2. The Caribbean Islands—Plato describes these two continents as lying at opposite ends of a navigable path of islands.

3. Argentina's Mesopotamian plain—Plato identifies the world's only rectangular plain defined by a 10,000-stadium long navigable waterway lying on one of the continents.

4. Plato correctly claims that this unique plain's climate is conducive to biannual harvests.

5. Plato provides a Mediterranean traveler's perspective of South America, whereby a traveler from the Mediterranean would initially make landfall along the continent's northeastern coast and witness a precipitous coastline extending southward until arriving at a flat even plain midway down the side of the continent.

6. Plato accurately notes that the mountains enclose the plain except where they open up toward the sea in the south.

7. He specifies a very restrictive 14.5-mile maximum distance between this unique plain and the sea; the distance between Argentina's Mesopotamian plain and the Rio de la Plata lies within this limit.

with an article in the German scientific journal *Naturwissenschaften*.[51] Balabanova had been researching drug use among the ancients, performing forensic analysis on nine Egyptian mummies that dated as far back as the 11th century BCE. Much to her surprise she found the mummies contained high concentrations of cocaine and nicotine, products of the coca and tobacco plants, which are indigenous to the Americas but not the African continent.

The discovery of these drugs in multiple Egyptian mummies

[51] *Naturwissenschaften* 79:358

dating from 1070 BCE to 395 CE suggests that the Egyptians engaged in transoceanic trade with ancient inhabitants of the Americas. This contradicts the views of today's anthropologists, who maintain that these drugs could not have found their way to Egypt until after Christopher Columbus' discovery of the Americas in 1492 CE. Critics immediately went on the defensive and refuted the findings with claims that the cocaine and nicotine were most likely external contaminates introduced after the mummies had been pulled from their place of rest, possibly in the form of archaeologists or museum curators who used cocaine and tobacco.

Balabanova, however, was no amateur in the field of toxicology. In fact, she pioneered methods used today for accurately determining drug use by individuals both living and dead. The methods she used are identical to drug testing techniques used today that decide the fate of many employees and determine conclusively if the deceased ingested drugs prior to death. One of those procedures included rinsing the hair shaft in advance of testing to verify the drug is *within* the hair and not a postmortem contaminant coating its exterior. Additionally, Balabanova's tests were replicated by others—F. Parsche and A.G. Nerlich, in the *Journal of Analytical Chemistry* 352:380-384—who also arrived at the same conclusion. The mummified Egyptians had indeed inhaled or ingested the drugs while living.

While critics are still grappling for possible—and thus far unproven—alternative explanations compatible with their unwavering view of history, Balabanova's findings remain the strongest line of evidence for transoceanic contact between ancient Egypt and the Americas. Plato's ancient Egyptian tale of a foreign power dwelling in a land matching the description of the Americas would seem to corroborate her findings.

The War That Was

Getting Plato's account to gel with our currently established historical view is complicated. There is scant historical documentation beyond Plato's dialogues and no apparent archaeological support.

In Plato's dialogue, Critias begins to delve into the details of the ancient war waged between the people of Atlantis and the peoples of the Mediterranean, claiming it occurred 9,000 years prior to Solon's journey to Egypt. Based on the approximate period of Solon's visit, this places the war circa 9600 BCE:

> *Let me begin by observing first of all, that nine thousand was the sum of years which had elapsed since the war which was said to have taken place between those who dwelt outside the Pillars of Heracles and all who dwelt within them; this war I am going to describe. Of the combatants on the one side, the city of Athens was reported to have been the leader and to have fought out the war; the combatants on the other side were commanded by the kings of Atlantis.*[52]

This 9,000-year period however, seems to be an erroneous recollection made by Critias. As we will find, Solon likely never specified the actual time period the war occurred. The only periods the Egyptians appear to have provided Solon were their beliefs that (a) ancient ancestors of the Athenians rose up around 9,600 BCE and (b) the Egyptian empire arose 1,000 years later.

Earlier in Timaeus, when Critias is reciting Solon's account, he provides the Egyptian priest's own words. The priest specifically states that 9,000 years in the past marks the year when the seed of Athenian civilization first appeared on Earth:

> *She founded your city a thousand years before ours, receiving from the Earth and Hephaestus the seed of your race, and afterwards she founded ours, of which the constitution is recorded in our sacred registers to be eight thousand years old. As touching your citizens of nine thousand years ago, I will briefly inform you of their laws and of their most famous action; the exact particulars of the whole we will hereafter go through at our leisure in the sacred registers themselves.*[53]

> *Yours first by the space of a thousand years, when she had received the seed of you from Ge and Hephaestus, and after that ours. And*

---

[52] *Critias* 108e

[53] *Timaeus* 23d, e

> *the duration of our civilization as set down in our sacred writings is 8,000 years. Of the citizens, then, who lived 9,000 years ago...*[54]

According to the priest, Egyptian civilization would not exist until 1,000 years after the rise of the ancient Athenians, circa 8600 BCE. This is where we first see the conflict in dates. The Egyptian priest goes on to say that the Atlantean force rose up not only against the Athenians but against the Egyptians as well:

> *This vast power, gathered into one, endeavoured to subdue at a blow our country and yours and the whole of the region within the straits.*[55]

It is problematic for the Atlanteans to rise up in 9600 BCE against an Egyptian empire that did not exist until 8600 BCE. What makes this even more problematic is that the Egyptians believed that all other civilizations, including the Atlantean empire, would not have existed until after 8600 BCE.

At the time Solon traveled to Egypt, there had been a very recent development in the Egyptian view of history. The Egyptians had long been of the mind that they were the most ancient of civilizations, but an experiment carried out decades before Solon's visit had determined that there was one, and only one, civilization that preceded their own. This new worldview appears to have influenced the priest's statement to Solon. According to the Greek historian Herodotus:

> *Now the Egyptians, before the reign of their king Psammetichus (664 - 610 BCE), believed themselves to be the most ancient of mankind. Since Psammetichus, however, made an attempt to discover who were actually the primitive race, they have been of opinion that while they surpass all other nations, the Phrygians surpass them in antiquity.*[56]

Like Solon, Herodotus had traveled to Egypt and received from their priests a similar revelation of a more ancient civilization, including

---

[54] *Timaeus* 23d, e; Bury translation

[55] *Timaeus* 25b

[56] Herodotus, *The Histories* 2.2, Translation by George Rawlinson.

the details of the not-so-scientific method leading to their discovery of this fact.

The details of this experiment are as follows: Egyptian King Psammetichus placed two young children in the care of a herdsman who was charged with the task of keeping them isolated from all people—and more importantly, all human speech. Psammetichus theorized that the first discernible word spoken naturally by the children would somehow be linked to the oldest of languages and therefore, by extension, the oldest of civilizations.

Once the children were of an age to speak, the first word out of their mouths was *bekos*. Upon investigation, it was found to be the Phrygian word for bread. And so it was from that point on the Egyptians determined one civilization, the Phrygians, were of more ancient origins than themselves.

We can question the methodology but this confirms Solon's account of the Egyptian view that their civilization preceded or "surpassed all other nations" with the exception of one. Combining both Solon's and Herodotus' accounts, the Egyptians believed that only one civilization came into existence circa 9600 BCE, 1,000 years prior to the Egyptian civilization. An unspecified time after the rise of the Egyptian nation, circa 8600 BCE, the Atlanteans arose. This rules out Critias' errant assumption of a 9600 BCE invasion of the Mediterranean; it is simply too early to be possible as it predates the existence of two of the three main combatants.

While Herodotus identifies this most ancient civilization as Phrygia, Critias clearly identifies Athens as the city that led the fight against the Atlanteans (Critias 108e). However, when Critias recites the priest's account, the words are actually tempered, stipulating that it was the city that "now" is Athens. This leaves open the possibility that it was not the Athenians themselves that fought the battle but a more ancient inhabitant of the region:

> *For there was a time, Solon, before the great deluge of all, when the city which now is Athens was first in war and in every way the best governed of all cities, is said to have performed the noblest deeds*

and to have had the fairest constitution of any of which tradition tells, under the face of heaven.[57]

If we were to directly reconcile the tale Solon received from the priest in Sais with the tale that Herodotus received from the priests of Vulcan in the city of Memphis, we might conclude that the Phrygians populated the ancient site of Athens and were the heroic warriors that fended off the Atlanteans.

The Phrygians, however, are an ancient people who dwelt in Anatolia, a region occupying much of modern Turkey. According to Herodotus, their ancestors the Bryges dwelt in the southern Balkans, placing them much closer to Athens.[58] But there is still no historical evidence that they ever moved south to occupy any portion of Greece.

There is little room to argue that the Egyptians clearly identified the Phrygians as the oldest civilization; the *bekos* connection in Herodotus' historical account is clear. Yet it is also clear that the priest in Solon's account created a very loose relationship between the modern Athenians and an ancient ancestor.

The priest explained to Solon that many catastrophic events had occurred in the past wiping away more advanced civilizations. These civilizations had established written language and history that the civilizations replacing them would not inherit:

> *Whereas just when you and other nations are beginning to be provided with letters and the other requisites of civilized life, after the usual interval, the stream from heaven, like a pestilence, comes pouring down, and leaves only those of you who are destitute of letters and education; and so you have to begin all over again like children, and know nothing of what happened in ancient times, either among us or among yourselves.*[59]

The priest even goes on to inform Solon that the Athenians were direct descendants of the extremely few remnants of the noble race that battled the Atlanteans:

---

[57] *Timaeus* 23c
[58] Herodotus, *The Histories* 7.73
[59] *Timaeus* 23a, b

> *As for those genealogies of yours which you just now recounted to us, Solon, they are no better than the tales of children. In the first place you remember a single deluge only, but there were many previous ones; in the next place, you do not know that there formerly dwelt in your land the fairest and noblest race of men which ever lived, and that you and your whole city are descended from a small seed or remnant of them which survived. And this was unknown to you, because, for many generations, the survivors of that destruction died, leaving no written word.*[60]

That the Egyptian priest claims that the Athenians "descended from a small seed or remnant of them which survived" provides a direct ancestral link to the Athenians. Assuming that both Solon and Herodotus related information from the Egyptian priests accurately, the Egyptians believed that the Phrygians and Athenians shared a linked lineage. Based on the Egyptian method of determining the most ancient civilization, language is likely the link that led them to associate the Phrygians with the Athenians. This link between the two languages is recognized by many today and Plato also recognized there were shared similarities between the two.[61]

Realistically, we may never know the true identity of the Mediterranean warriors who fought the war against the Atlantean invaders. But we come away with the knowledge that the ancient Athenians we know were not the Mediterranean combatants. Rather, a more ancient civilization, an ancestral people, is believed to have dwelt in the region and fought against the Atlanteans. The date of that battle is neither 9600 nor 8600 BCE, but an unspecified date after the rise of the Egyptian civilization and prior to the rise of the Athenian nation of which Solon was a part.

Those Who Dwelt Outside

There is similar difficulty in identifying the foreign combatants

---

[60] *Timaeus* 23b, c

[61] *Cratylus* 410a

hailing from the Atlantic. Aside from a very accurate description of the Americas, there appears to be no archaeological evidence supporting the existence of this ancient sea power, though Plato does provide a clue about their surviving descendants:

> *The eldest, who was the first king, [Poseidon] named Atlas, and after him the whole island and the ocean were called Atlantic. To his twin brother, who was born after him, and obtained as his lot the extremity of the island towards the Pillars of Heracles, facing the country which is now called the region of Gades in that part of the world, he gave the name which in the Hellenic language is Eumelus, in the language of the country which is named after him, Gadeirus. Of the second pair of twins he called one Ampheres, and the other Evaemon. To the elder of the third pair of twins he gave the name Mneseus, and Autochthon to the one who followed him. Of the fourth pair of twins he called the elder Elasippus, and the younger Mestor. And of the fifth pair he gave to the elder the name of Azaes, and to the younger that of Diaprepes. All these and their descendants for many generations were the inhabitants and rulers of divers islands in the open sea; and also, as has been already said, they held sway in our direction over the country within the Pillars as far as Egypt and Tyrrhenia.*[62]

So while the capital of Atlantis links to Atlas and the central region of South America, the Atlantean combatants also appear to have inhabited a region of the continent named after Atlas' twin brother Gadeirus as well as the many islands in the open sea bestowed upon Poseidon's remaining eight sons. All these are said to have "held sway" over the country within the Pillars (Fig. 64).

According to Plato's account, Atlas' twin Gadeirus was allotted the region of the island continent that lay toward the Pillars of Heracles. Assuming South America is Atlantis, this would place it in the northeast portion of the continent. Gadeirus would likely be located in an area with exceptional access to the sea, following the example of the capital city allotted to Atlas. The mouth of the Amazon River fits well, providing a desirable region worthy of the second son. The region features both a

---

[62] *Critias* 114a-c

Figure 64. *Map positing the extent of Atlantean influence at its peak and the possible location of Gadeirus in the Amazon region. "In this island of Atlantis there was a great and wonderful empire which had rule over the whole island and several others, and over parts of the [opposite] continent, and, furthermore, the men of Atlantis had subjected the parts of Libya within the columns of Heracles as far as Egypt, and of Europe as far as Tyrrhenia." - Tim 25a*

Gadeirus was claimed to be the portion of the island continent that lay toward the Pillars of Heracles, facing Gades. Solon inserted this information clearly believing the Phoenician city of Gades to be Gadeirus' namesake, linking the Phoenicians to the Atlanteans.

productive agricultural plain and low-lying coastal access for seagoing vessels, very similar to the plain surrounding the posited capital city allotted to the number-one son, Atlas.

In the case of Gadeirus, Solon provides the only link between the ancient Atlanteans and its remnants still in existence in his day. It is no coincidence that Solon goes out of his way to provide Eumelus' non-Hellenic name Gadeirus, the only son of Poseidon's ten for whom he includes the name in its original language. In similar fashion, Solon makes a point of including the city of Gades as a directional guide for the location of Gadeirus when the Pillars of Heracles, located in the same vicinity, would have sufficed. Solon clearly wished to make the reader aware that there was a link between Gadeirus and Gades, a connection that would have been lost had he only provided the reader the Hellenic name Eumelus.

The only possible reason for Solon including this link was a belief that there were ancestral ties between the Atlanteans and the Phoenicians, who were the founders of Gadeirus' namesake, the ancient city of Gades, but there is currently no means of verifying this ancestral link. The Phoenicians are believed to be of Semitic lineage, but no historical record exists outside of Plato's dialogues that provides any clue as to the Atlantean lineage.

However, the Phoenicians are in many ways the ideal suspect. They were the most skilled seafarers and ship builders of their day. This combination of highly seaworthy vessels and advanced seamanship allowed them to undertake extraordinary voyages that exceeded the capabilities of their contemporaries. So exceptional were these feats and their knowledge of the seas that at least one ancient Greek historian voiced skepticism toward a claim made regarding one of these voyages.

Herodotus relates how Egyptian King Necho II, upon abandoning the construction of a canal that would have connected the Mediterranean Sea to the Red Sea, dispatched a fleet of Phoenician vessels on a three-year voyage to circumnavigate the African continent.[63] They set out from a location in the Red Sea sailing down along the eastern African coast, stopping ashore for a time to sow and harvest a crop of corn

---

[63] Herodotus, *The Histories* 4.42, 43

for the remainder of their journey. The final year of the voyage, they entered the Pillars of Heracles and made their way to the mouth of the Nile, marking the voyage's completion.

The claim for which Herodotus expressed doubt was "that in sailing round Libya [the Phoenician sailors] had the sun upon their right hand." Had the Greeks been similarly adventurous, they would have known that this claim was actually true, as sailing west around the tip of Africa placed the sun to the north, or off the right hand of the westward-facing sailors.

Herodotus also claims that the Carthaginians, who were of Phoenician lineage, made this voyage, but only recounts one voyage that was partially completed. Pliny the Elder confirms they had made the voyage when he relates that Hanno the Navigator, a Carthaginian explorer, circumnavigated the African continent sailing "round about from Gades to the utmost bounds and lands-end of Arabia."[64]

According to the Hebrew Bible, King Solomon also enlisted the Phoenicians' services.[65] Hiram I, the Phoenician king of Tyre, formed an alliance with Solomon. During this time, a fleet of Phoenician trade ships manned by Hiram's men made regular voyages to Ophir carrying "gold, silver, ivory, and apes and baboons" back to Israel every three years. It is unclear where Ophir is located today, but the three-year interval that matches Herodotus' account for circumnavigating Africa suggests a journey of comparatively long distance.

The idea that the Phoenicians were the descendants of Atlantis may have some merit if we consider the possibility that they retained advanced seafaring skills from their predecessors. Another skill they seem to have retained is canal digging, as we have seen an impressive array of manmade canals one hundred feet in width crisscrossed the vast plain and defined at least a portion of the waterway encircling the plain. The canal that Necho II would eventually abandon was dug by various nations, but it is said that the Phoenicians were far more efficient and successful in digging their assigned portion.

---

[64] Pliny, *Natural History* 2.169

[65] 1 Kings 10; 2 Chron. 9

> *All the other nations, therefore, except the Phoenicians, had double labour; for the sides of the trench fell in continually, as could not but happen, since they made the width no greater at the top than it was required to be at the bottom. But the Phoenicians showed in this the skill which they are wont to exhibit in all their undertakings. For in the portion of the work which was allotted to them they began by making the trench at the top twice as wide as the prescribed measure, and then as they dug downwards approached the sides nearer and nearer together, so that when they reached the bottom their part of the work was of the same width as the rest.*[66]

Over the last hundred years or so, many have come forward with Phoenician inscriptions claimed to have been discovered in the Americas. Unfortunately, all have either been proven to be hoaxes or remain highly suspect. These hoaxes will only make it more difficult to accept any future inscription as genuine. It will require a find far more extensive than inscriptions alone to prove that the Phoenicians were present in the Americas.

Similarly, true validation of an ancient Atlantean civilization will likely only come with the discovery of their capital city, an ancient city with a concentric design in a geographic location that complies with the specifications detailed in the Atlantis account. Currently archaeology in the Americas and in the near vicinity of the Paraná find no evidence supporting the existence of an ancient advanced seafaring civilization, supporting a consensus among anthropologists that no such civilization ever existed.

The chapters that follow will consider more closely the veracity of many established beliefs. Theories that have remained unchallenged for decades and even centuries may be long overdue for a fresh look. These upcoming chapters also mark a transitional point where we move from gleaning truth in fringe beliefs to exposing fallacy in scientific consensus.

---

[66] Herodotus, *The Histories* 7.23

PART II

MODERN MAPS

Paradigms and Puzzles

CHAPTER 10

PARADIGM LOST

Having postulated the possible home of Atlantis in South America, we have placed a capable ancient seafaring culture in the closest possible proximity to Antarctica. Deep in the realm of a highly hypothetical lost history, this civilization would be the most likely source for an ancient Antarctica map. This same civilization appears to have passed along the earliest mention of two continents beyond the Mediterranean and details of the Americas that include an accurate description of Argentina's Mesopotamian plain and the Caribbean Islands.

Of course, there are many barriers to proving this theory true. They include how a widespread ancient civilization became completely lost in time and managed to avoid the picks and shovels of a curious modern world. Even more problematic is how an ancient Antarctic map merely thousands of years old could accurately portray the continent devoid of the ice sheet believed to have encapsulated it for tens of millions of years.

According to Plato, the Egyptian priest and source of the Atlantis account provided the reason for the lost history. He claimed it was due to a series of floods in the Mediterranean that included a "great deluge of all," which wiped out both the Atlanteans and most of those that knew of the Atlantean empire.

> *In the first place you remember a single deluge only, but there were many previous ones; in the next place, you do not know that there formerly dwelt in your land the fairest and noblest race of men which ever lived, and that you and your whole city are descended from a small seed or remnant of them which survived. And this was unknown to you, because, for many generations, the survivors of that destruction died, leaving no written word.*[67]

The noble ancestors of the Athenians who triumphed over the Atlantean force were also decimated by flooding. The account mentions simultaneous earthquakes, suggesting tsunamis were a main source of the flooding. Not only did the earthquakes and tsunamis wreak havoc on the inhabitants of the Mediterranean, but Atlantis was also subjected to the same desolating effects. This ties into Solon's suggestion that the Phoenicians and Athenians were remnants of two separate lost civilizations that met their catastrophic demise toward the end of a great war:

> *There occurred violent earthquakes and floods; in a single day and night of misfortune all your warlike men in a body sank into the earth, and the island of Atlantis in like manner disappeared in the depths of the sea.*[68]

This clearly describes a global cataclysmic event. Because of the closed nature of the Mediterranean, it would be difficult for a tsunami from the Atlantic to maintain much force passing through the Strait of Gibraltar and wreaking any substantial havoc on the Mediterranean, or vice versa. The devastation therefore suggests a global cataclysm consisting of quakes and tsunamis occurring simultaneously inside and outside the Mediterranean.

Consider also the effects that a global cataclysm of this nature might have on ocean currents. If it led to major shifts in the currents, sudden cooling in the southern hemisphere may have initiated rapid ice accumulation in Antarctica, supporting the possible legitimacy of a pre-glacial map of the continent. Of course, we would still need to

---

[67] *Timaeus* 23b, c
[68] Ibid., 25c, d

overcome the disparity between an ice cap believed millions of years old versus a few thousand.

Obviously, forming a reasonable hypothesis to support a global cataclysmic event of this scale would be an extremely difficult challenge. Initially I was almost certain that this would prove to be an unsuccessful venture but pressing on in this direction seemed the most logical way to proceed.

I began searching for signs of a global cataclysmic event by studying modern topographic relief maps, which detail Earth's surface, and bathymetric relief maps, which chart the depths and terrain of the ocean floor. One of the first things that stood out as I reviewed bathymetric maps was that the seafloor is strewn with many complex patterns of ancient activity. When these patterns are analyzed, it becomes clear that the current concept of seafloor spreading is an undeniable reality.

Evidence of seafloor spreading was first produced in 1961, when scientists researching patterns of magnetic striping along the Mid-Atlantic Ridge found that the patterns not only paralleled the ridge but did so in a mirroring pattern from one side to the other. They surmised that new crustal material formed at the central ridge recorded periodic reversals in Earth's magnetic field. Seafloor crust transported out and away from the central ridge in both directions made way for a continuous feed of new molten seafloor that cooled and hardened, acting as a record of multiple periods of magnetic pole reversals (Fig. 65 Inset).

Not only do we find magnetic striping radiating mirrored patterns out from these expansion ridges; physical deformations originating at the ridge are also conveyed outward, providing further proof that these plates are the product of seafloor expansion. For instance, two uniquely curved troughs in the South Pacific extending vertically down from the Menard Fracture Zone (MeFZ) similarly dip down over 20,000 feet below sea level, 4,000 feet below the surrounding seafloor (Fig. 65). These two troughs lie roughly 850 miles to the west and 950 miles to the east of the East Pacific Ridge. There is no other reasonable explanation for the existence of these unique twin features to have occurred similar distances from the central ridge outside of their having been the product of a shared tectonic event originating at a previously shared point, the

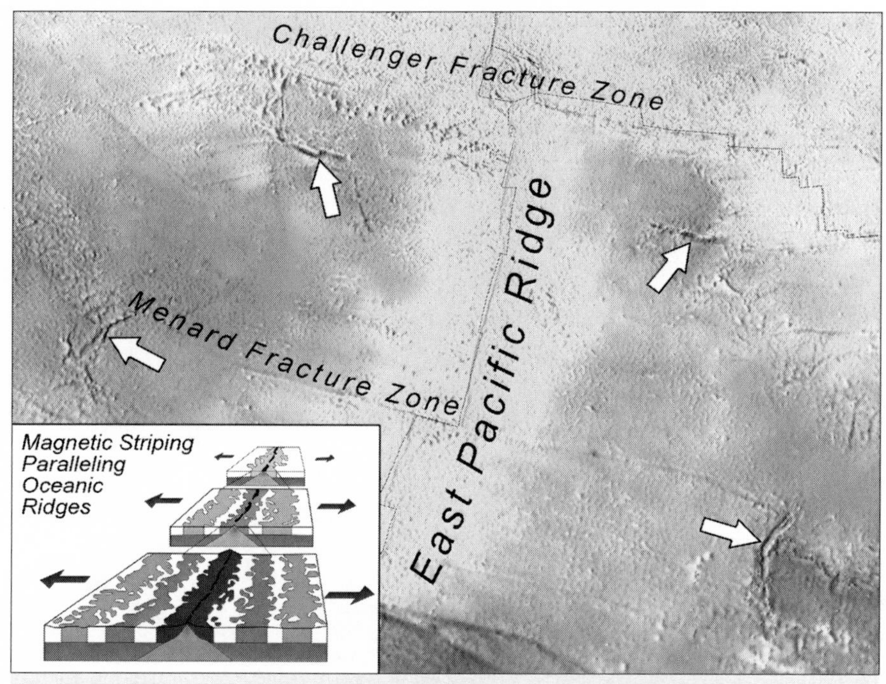

Figure 65. *Magnetic striping (inset) paralleling an oceanic ridge provides proof that the pattern emanates from a central point, a mid-oceanic ridge. In the South Pacific, mirrored troughs both lateral and vertical lie equidistant from the East Pacific Ridge. There is no possibility that these near-identical features formed independently over a thousand miles apart, as they share the same relative positions to nearby fracture zones and are equidistant from the East Pacific Ridge. The only reasonable conclusion is that these troughs were produced by tectonic activity occurring at the central ridge and were then transported away from the ridge as new seafloor crust was generated.*

East Pacific Ridge. Ridge buckling or some other heightened ridge activity temporarily dropped the new molten seafloor ridge by 4,000 feet for a short instance, then conveyed the crustal blemishes outward toward the east and west simultaneously at the same relative rate of speed. These geographic features were not formed by separate events occurring 1,800 miles apart. Evidence that these shared a common origin at the central ridge lies in the consistent occurrence of mirrored features across expansion ridges.

More geographic mirroring can be seen in the South Pacific just north of the MeFZ where two lateral troughs, similarly dipping down to

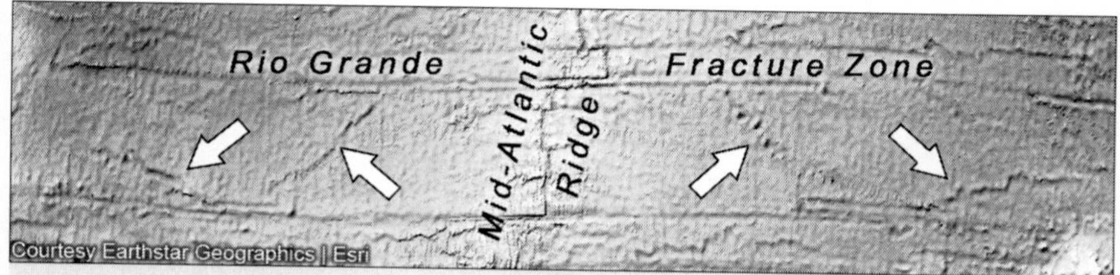

Figure 66. *Mirrored troughs and ridges are not limited to the Pacific. In the Atlantic we can also find evidence of seafloor spreading along the Mid-Atlantic Ridge. Lying just below the Rio Grande Fracture Zone at the 29th parallel south, we find mirrored trough-like features angling downward away from the Mid-Atlantic Ridge and other mirrored troughs following a shallow rise from the fracture zone below.*

7,000 feet below the surrounding seafloor, sit roughly equidistant from the East Pacific Ridge at about 500 miles.

Vertical troughs and ridges identically mirror across the Mid-Atlantic Ridge as well. One significant pattern of mirroring can be found just beneath the Rio Grande Fracture Zone (RGFZ; see Figure 66). Here we find one mirrored set of troughs dropping down and away from the expansion ridge, while another set below rises away much more gradually.

Generation of mirrored features across a central expansion ridge can be demonstrated by the use of two rolls of plotter paper configured as illustrated in Figure 67. The two rolls are placed side by side and configured with an inner dispensing roller and an outer take-up roller. Where the two rolls of plotter paper come together in the center, two plotting devices are moved up and down the crease, simulating random tectonic activity at the central expansion ridge. As the two rolls of paper are simultaneously reeled onto their corresponding outer take-up rollers, they transport a record of central activity outward onto a continuous feed of freshly revealed material. In this particular instance, the plotter's interaction with the paper's outward movement has recreated the patterns of mirrored troughs and ridges found lying below the RGFZ in the Atlantic, as seen in Figure 66.

The evidence clearly demonstrates that these features originated at a common centralized point, an oceanic expansion ridge, and were transported away from the ridge as the seafloor expanded. Earth scientists in the 1960s had this correct early on but were left puzzled

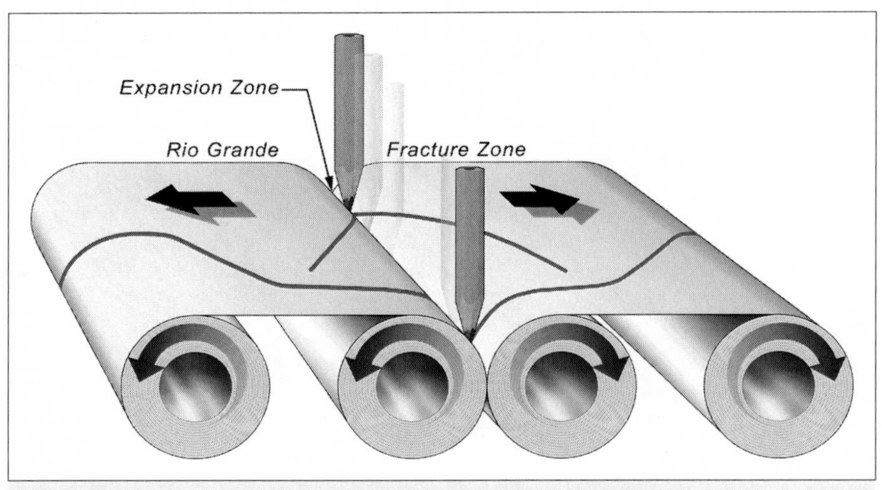

Figure 67. *Replication of mirroring effect found beneath the Atlantic's Rio Grande fracture zone using adjoining rolls of plotter paper. Plotting devices mimicking tectonic activity at a central expansion ridge are vertically active. Meanwhile the writing medium, representative of the generation of new seafloor crust flowing out from the expansion zone, is simultaneously etched with mirrored lines mimicking those seen in figure 66.*

by the unresolved issue of excess seafloor crust. If new crust were continuously being created and transported outward from these mid-ocean ridges, the older excess crust lying furthest from the ridge would either (a) find itself folding in upon itself, creating mountain ranges rivaling the Himalayas, or (b) the earth would have to continually expand to accommodate the never-ending supply of new crust.

Because the continental coasts do not exhibit adequate accumulation of mountainous folds of ancient seafloor crust, the theory that Earth expands to accommodate all new seafloor crust remained a fairly popular and legitimate theory well into the 1950s.

From the 1960s on, however, a new consensus formed with the discovery of deep oceanic trenches in the Western Pacific. The new theory, which would come to be known as plate tectonics, advocated a static, non-expanding Earth where seafloor plates are transported from an expansion ridge toward and into these deep trenches by large powerful convection cells within Earth's superheated fluid mantle. These trenches in turn mark subduction zones where ancient excess crust is diverted back into the Earth's mantle to be melted down into magma. This creates a cycle of crustal destruction at the trenches and regeneration at expansion ridges. This leaves the planet's size intact, conforming to a planet that currently exhibits no clear signs of spherical expansion.

As scientists continued to develop the theory of plate tectonics, they found that their theory was confirmed by radiometric dating. Drilled core samples from the seafloor exhibited gradually older crustal ages extending from the central ridge outward, with some of Earth's most ancient crust lying near the trenches where it was believed to subduct back into Earth's mantle.

The further I studied maps of the planet, the more I began to believe that there was no reason to doubt the soundness of the plate tectonics theory, which conformed to the widely accepted uniformitarian view. This view assumes that Earth's plates have and always will move at their current slow rate of speed. Catastrophism is just the opposite of uniformitarianism and would have proved better suited to explain a global cataclysmic event.

Uncertain where to turn next, I decided to address a seemingly harmless curiosity that had drawn my focus for decades. I had no idea at the time that this simple observation I had carried with me from my youth would potentially be the key to unraveling a modern paradigm and establishing a new earth dynamic. In fact, I began researching this curious geographic feature believing it further supported the basic dynamic of plate tectonics.

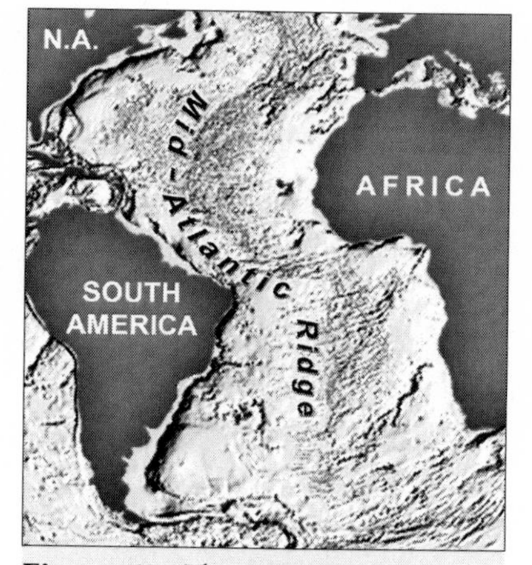

Figure 68. *The Americas and African continent conform to each other, and both conform to the central mid-Atlantic expansion ridge.*

When I was far younger and first became aware of the theory of plate tectonics and, more specifically, continental drift, I was taken in by the stunning proof of continental drift as seen in the near perfect fit of continents across the Atlantic (Fig. 68). It was not long before my fascination with maps led me to a portion of land in the Pacific that exhibited this same unmistakable conformance. It was a place I had always been drawn to whenever I found myself studying maps of the Pacific but had never had the inclination or the resources to research until now: the Kamchatka Peninsula.

Kamchatka: Window to a New World

As scientists continued to study trenches, believing them to be subduction zones, they began looking for an explanation for the paralleling paths of volcanic islands that formed along the adjoining overriding crust, paths known as island arcs. It was determined that the movement of the descending or subducting plate beneath an opposing plate generated immense friction, leading to volcanic activity or magmatic upwelling that melted through the overriding plate. This superheated material would eventually breach the surface of the ocean floor, forming a series of volcanic islands.

The Kamchatka Peninsula, which extends off Russia's eastern coast out into the Pacific (Fig. 69), is one example. It is part of the Kuril-Kamchatka Arc, an island arc believed to have been formed as the Pacific Plate subducted under the Okhotsk seafloor. As the Pacific crust subducted beneath the Okhotsk Plate beyond the earth's lithosphere and down into the asthenosphere,[69] the friction between these two plates is

Figure 69. *The Kamchatka Peninsula, which extends from the eastern coast of Russia.*

[69] The lithosphere comprises Earth's crust and the rigid outermost layer of the mantle lying directly beneath it. The asthenosphere is the layer of Earth's mantle lying below the lithosphere believed to be partially molten and ductile in nature.

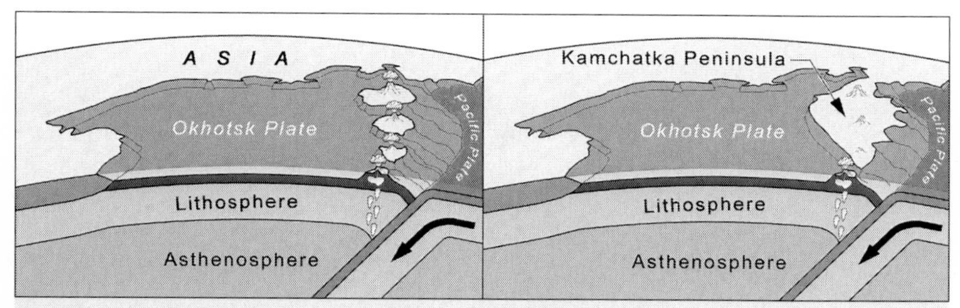

Figure 70. CURRENT THEORY: *According to plate tectonics, the Kamchatka Peninsula and its associated island arc were formed as the Pacific Plate subducted beneath the Okhotsk Plate. Friction between the two plates is believed to have generated volcanic activity, creating a series of islands. Islands in the north merged together and expanded to the Russian coastline to form the Kamchatka Peninsula.*

believed to not only generate frequent and often intense earthquakes, but cause partial melting of the subducting Pacific crust. This melted crust subsequently melts through the overriding Okhotsk Plate and erupts through the seafloor, forming volcanic islands. It is believed that the Kuril-Kamchatka chain of islands was formed in this fashion (Fig. 70 left).

I had long believed that islands within island arcs were formed in this manner and saw little wrong with the theory. What I was not aware of is that geologists also maintain that the Kamchatka Peninsula shares similar origins. According to geologists, Kamchatka initially rose up as a series of volcanoes, much like the remainder of the island arc, but increased activity saw the islands grow larger and expand together, eventually flowing together into one big composite island. Over time this large island expanded further to the north until it connected with the Asian continent and became the peninsula it is today (Fig. 70 right).

I believe that Kamchatka's true genesis is far less complicated, as made clear by the evidence that follows. Based on the conformance of continents straddling the Atlantic, I had expected scientists to have similarly recognized Kamchatka's near-perfect conformance with the adjacent Asian coastline. As can be seen in Figure 71, the adjacent Asian coastline accommodates the Kamchatka peninsula quite well, conforming to Kamchatka's signature southern point and humped western coast. Adding further confirmation is a very small and unique

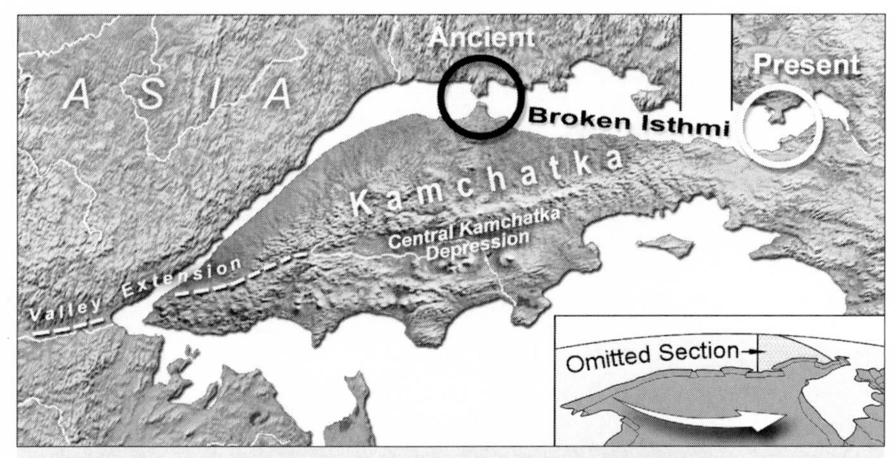

Figure 71. *A merged image of Kamchatka pocketed back into the Asian coast, where two lone coastal points align (top center circle) revealing an ancient isthmus. Demonstrating that this is not coincidence, the image includes a similar broken isthmus exactly as it exists today further up the coast of Kamchatka (top right circle). Under the current theory, both of these alignments would have to be considered pure chance.*

The Central Kamchatka Depression and the valley extending from it align with a well-defined valley extending into the Asian mainland, further confirming the true origin of the Kamchatka Peninsula.

coastal point located on Kamchatka's western coast, which is squared off at its furthest extremity. There is a lone matching coastal point on the Asian coast that lines up well when Kamchatka is positioned back into place. This refit reveals that the two coastal points were at one time joined together, forming a land bridge or isthmus that broke apart as Kamchatka separated from the mainland.

Is there a chance that this alignment of lone coastal points is only coincidental? Not likely. There are two other outward projecting coastal points in the region that similarly align. There is no need to perform any map manipulation to imagine their alignment as these two coastal points remain aligned to this day and lie a mere 18 miles short of forming an isthmus (Fig. 71 top right). Like the other two coastal points, I believe these two were once merged together, forming an isthmus that fractured relatively recently as Kamchatka continued to break free of the Asian mainland.

The refit also exposes Kamchatka's dual parallel mountain ranges as a consequence of crustal folding occurring while the peninsula was still joined to the Asian mainland. This is evidenced by the deep valley lying between Kamchatka's two ranges aligning with a similarly sandwiched deep valley extending out onto the Asian mainland when Kamchatka is positioned back along the coast. Confined to the standard subduction model, earth scientists continue to struggle to reconcile the origins of Kamchatka's dual linear volcanic ranges that line this valley. Subduction and island arc formation might explain the trench-side range, but the existence of the secondary western range lying further inland has led some to consider a very complicated process of layered, colliding ridges. However, this new model provides a much more obvious and simplified explanation, making such notions appear far less credible.

Based on these observations, it is certain that instead of an isolated genesis erupting up from the seafloor hundreds of miles from the Asian coast and randomly coalescing into a landform conforming perfectly to the mainland, Kamchatka is a continental fragment that began its existence merged with the Asian continent.

Bays: Continental Ductile Fractures

In the interest of full disclosure, there is a sizable disparity between the total length of Kamchatka's western coastline and the adjacent continental coast that was edited out of the previous image for initial clarity. Kamchatka has a 950-mile coastline, while the accommodating Asian coast has a length of 1,300 miles, a difference of 350 miles visible in Figure 72's unmodified map of the region. Open gaps along the Asian coast in the form of two bays contribute to this discrepancy.

The spans of the two bays are 80 and 195 miles across, totaling 275 combined miles. These two bays account for most of the 350-mile disparity. The two bays demonstrate a basic phenomenon readily recognized in the study of fracture mechanics known as ductile fracturing.

Rigid materials can fracture or break apart with fragments displaying little to no sign of deformation. These breaks are known as brittle fractures. If you have ever broken a glass or a ceramic plate, you have witnessed this type of fracture. In many instances, the fragments

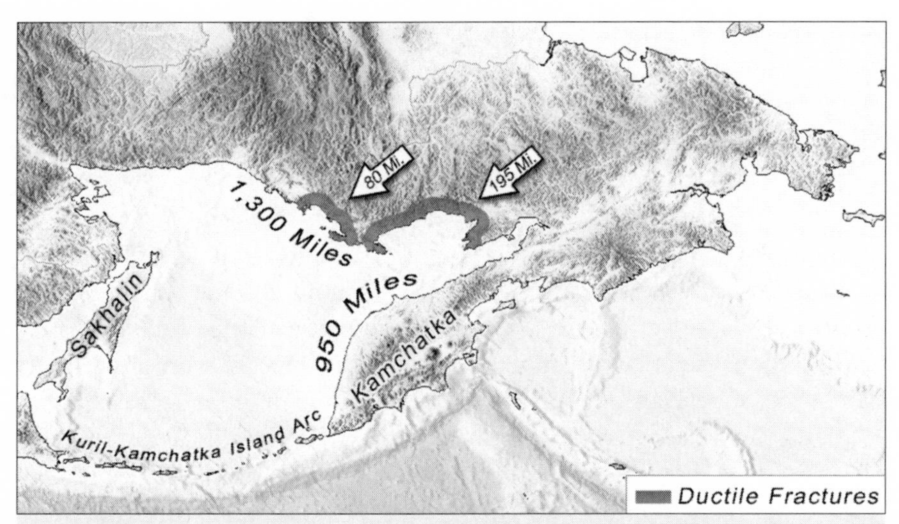

Figure 72. *Almost the entire 350-mile discrepancy between the Kamchatka western coast and the Asian coast can be found in the existence of two large bays, which measure a combined 275 miles between cusps.*

fit back together so well that the objects can be fully reassembled to their original form. Most of Kamchatka's break from Asia exhibits this type of fracturing with minimal deformation, allowing us to see its original near-perfect fit.

Meanwhile, material with the right ductility or plasticity—for instance, heated plastic or salt-water taffy—is susceptible to substantial deformation while fracturing. As the material fractures, stretching and thinning will cause one or more oval voids to open up in the midst of the fracture point. As the material is torn apart, the outlying edges of these oval voids tend to form outstretched wispy points or cusps. This is similar to what we are witnessing along the Asian coast.

Fracture mechanics is a field dedicated to the study of material properties when subjected to stress and cracking. Figure 73 demonstrates ductile fracturing as it occurs in soft sheet metal. It clearly exhibits similarly arced tearing with outstretched cusping, as we see in the two bays along the Asian coast.

It therefore appears that while the Kamchatka fracture is largely a brittle fracture where very little distortion occurs between mating coastlines, there are also instances of ductile fracturing or tearing. This is evidenced by these arc-shaped pockets or bays along the mainland coast, formed where more pliable portions of the continent succumbed to extensive stretching during fracturing. In other words, the same force that ripped Kamchatka clear of Asia also stretched and ripped open these bays along the Asian coast.

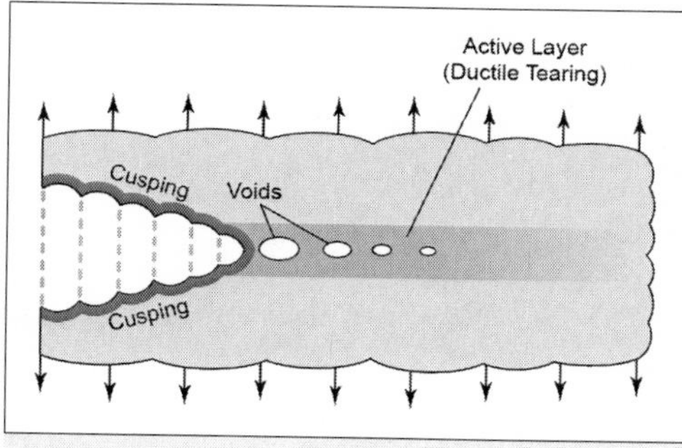

Figure 73. *Ductile fracturing as it occurs in sheet metal. The fractures exhibit similar arcing tears and outstretched cusps found in the continental crust in the Kamchatka region.*

Shortly after making this discovery, I noticed that there were numerous similar fractures prevalent throughout the region. As we will see, ductile fracturing, a basic aspect of materials properties, is an overwhelmingly common occurrence in continental crust. Though this potentially holds the keys to unlocking Earth's past, it seems to have gone mostly unnoticed by geologists.

Figure 74 highlights the many ductile fractures in the immediate vicinity of Kamchatka. Ductile arc AB on Kamchatka's eastern coast links back to points A1 and B1 forming an arc along the mainland, with B correlating to the northernmost tip of Sakhalin. Like the ductile sheet metal in Figure 73, these two arcs were once connected, forming an open circular ductile void prior to Kamchatka completely breaking away.

The remaining ductile arcs are of a slightly different nature and origin, occurring along Kamchatka's exposed eastern coast. They are similar to the two arcs northwest of Kamchatka already discussed but with cusps pointing outward. Again, the same forces that ultimately unseated Kamchatka from Asia exhibit themselves all along Kamchatka's eastern coast and extend northward, demonstrating a continuous strain on the continental plate. As a result, the whole of this northeastern region of Asia appears as if it is in the process of being ripped down and away from the Arctic while Kamchatka is left behind.

It seems likely that this fracturing process began when the Asian

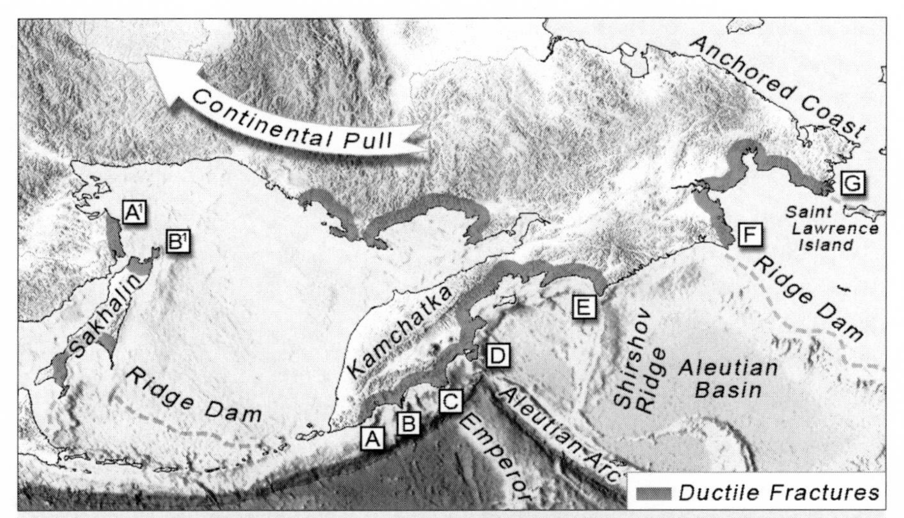

Figure 74. *The Kamchatka region is rife with ductile fractures. Not only have these fractures thus far gone unobserved, but it has gone unnoticed that the cusps of these fractures are linked to seafloor ridges. In almost every instance, the cusps are directly in line with their associated seafloor ridge. The exceptions are cusps A and B; these cusps aligned with A¹ and B¹ when Kamchatka was still embedded into the Asian coast. The two arcs formed a circular ductile void prior to Kamchatka's extraction from Asia.*

Another discovery relates to the revelation that some of these ridges act as dams for sediment deposits, forming continental shelves (southern tip of Sakhalin and point F). Saint Lawrence Island exposes a partially buried ridge in the Bering Sea that aligns with a northern cusp (G).

continent retreated away in a clockwise direction from the upper northeastern region around Saint Lawrence Island in Canada, which remained anchored to the Arctic. As the Asian continent began its retreat down and away with Kamchatka still pocketed in along its side, the initial stress was applied along the eastern coast of Kamchatka and northward. This coastal strain caused ductile fractures to snap open, forming a chain of bays along the edge of the continental plate where points B though G are located, while void AB began to open up as an internal void as well.

As we will see later, Kamchatka itself was the product of a preexisting brittle fracture that had not fractured entirely through. It was only with the application of ongoing stress along the coast that the peninsula began to extricate itself from the continent relatively suddenly.

As this continental fragment, largely separated but still pocketed into the Asian coast (reference Fig. 71) began to take on its peninsular form, the first of two outstretched ductile isthmi reached its breaking point and fractured apart near its center, launching the peninsula onward. Continuing on its way, the ongoing stress on the continent saw the Asian coast beyond the broken isthmus fracture, tearing the two large elongated bays into the mainland.

Finally, the continental fragment, continuing on its relative counterclockwise course, saw further separation of the peninsula with the one last ductile isthmus forming and eventually breaking in two. This brings us to the region's current geographic configuration.

While recognition of these ductile features helps paint a clear picture of Kamchatka's true genesis, it also reveals a previously unobserved geological correlation. Many of these fractures are linked to visible ridge formations extending directly from their outstretched cusps. Points C, D, and E clearly align with the Emperor seamount chain, the Aleutian Island Arc, and Shirshov Ridge respectively, but other cusps like point F and those at the southern tip of Sakhalin align with the leading edge of sediment deposits or continental shelves. It suggests that continental shelves may in many instances be the product of sediment deposits settling behind seafloor ridges that function as dams.

Continental shelves are currently believed to be the product of natural sedimentation having flowed mostly unrestricted off a continental mass out onto the open seafloor. However, in that case one would expect to find the sediment forming a smooth gradual descent to the seafloor, instead of the sudden abrupt drop characterizing the leading edge of most continental shelves. Identifying the front edge of these shelves as seafloor ridges covered with sedimentary overflow would seem to account for the sudden sharp descent to the seafloor.

Cusp G also appears to align with a seafloor ridge. It is not readily apparent, as sediment has overrun most of this particular ridge, but the cusp aligns with a linear Saint Lawrence Island. It is therefore likely the partial breaching of a buried ridge, as nearly all linear islands are associated with ridge formations, as seen in both the Aleutian and Kuril Ridges.

Major Ridges Consistently Signify Plate Boundaries

Geologists offer varying explanations for the origins of each of these regional ridges. However, I believe the consistency in their alignments, with outstretched cusps off ductile fractures, offers a far more consistent global view of the relationship between continental crust and seafloor crust creation. The first observation finds that seafloor crust is actually cemented to continental crust and reacts directly to continental plate movement. Where continental crust fractures and separates, the adjoining seafloor crust is likewise fractured and pulled along incrementally, exposing magma, which cools and creates new seafloor crust.

Figure 75 demonstrates the basic process by which the continental crust undergoes ductile fracturing and directly affects the adjoining seafloor crust. In the upper image, the underlying magma is exposed as a void opens within the fractured coast; it extends out into the adjacent seafloor as the continental crust pries the seafloor apart. The exposed magma subsequently cools and hardens into new seafloor crust. The ridges that define the outer edge of this new crustal material are therefore boundary ridges separating the newly generated seafloor crust from the surrounding seafloor crust. The material lying between these ridges forms a new seafloor crustal plate amidst the old, an expansion zone.

The bottom image in Figure 75 reveals how elementary this concept truly is. It equates a wooden frame to continental crust and the attached canvas to seafloor crust. Like the canvas, seafloor crust is attached and stretched between continental plates. When the frame or continental crust is fractured, it rips apart the attached material creating an outward void that continues to extend as the fracture in the frame or continental crust widens. The main difference between the canvas and the seafloor crust is that molten magma lies below the crust incrementally filling the void as it expands. Along the boundaries, bonding between old and new seafloor crust occurs slowly. This creates instability between the

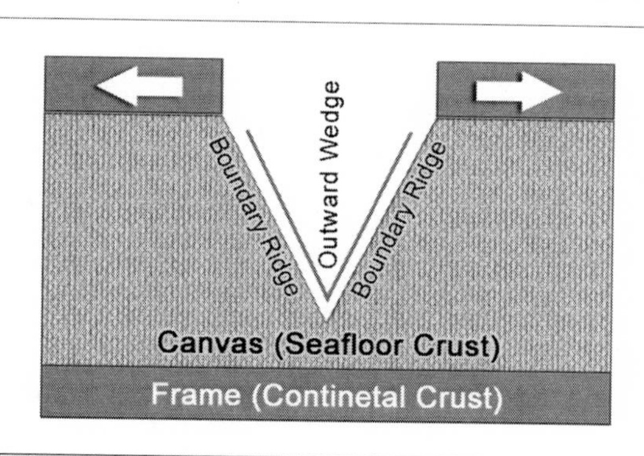

Figure 75. *The association of ductile fracture cusps with seafloor ridges suggests that seafloor crust is bonded to continental crust. Within this new dynamic, fracturing of continental crust along the coast rips open the adjoining seafloor crust with magma seeping between old and new plates forming boundary ridges. This dictates that the surface of our planet is much like a canvas frame (bottom illustration). If we were to break the frame, the attached canvas would respond by ripping outward into the canvas forming a V-shaped void. Likewise, when our planetary crust fractures along the coast, it rips a similar outward void into the attached seafloor crust.*

two crusts which allows magma to seep through forming boundary ridges. These ridges extend out from the cusps of the continental ductile fracture to the farthest end of the torn seafloor crust.

This link between ductile fractures and ridge boundaries brings us to the second observation—one that creates a potential dilemma for the plate tectonics theory. This dilemma is a ridge and cusp alignment. It involves the Emperor seamount chain, the upper extension of the Hawaiian-Emperor seamount chain; see cusp C in Figure 74.

Early on scientists believed, as we reestablish here, that volcanic ridges signified plate boundaries. Yet this ridge differs greatly. In the development of the plate tectonics theory, this unique ridge proved very puzzling to geologists in that it extends linearly thousands of miles from the coast of Asia, only to end abruptly in the middle of the Pacific Ocean at the Hawaiian Islands. In 1963, Canadian geophysicist and geologist J. Tuzo Wilson formulated a theory to explain the existence of this ridge and others like it across the globe, which would come to be known as the hotspot theory. According to this theory, as the Pacific

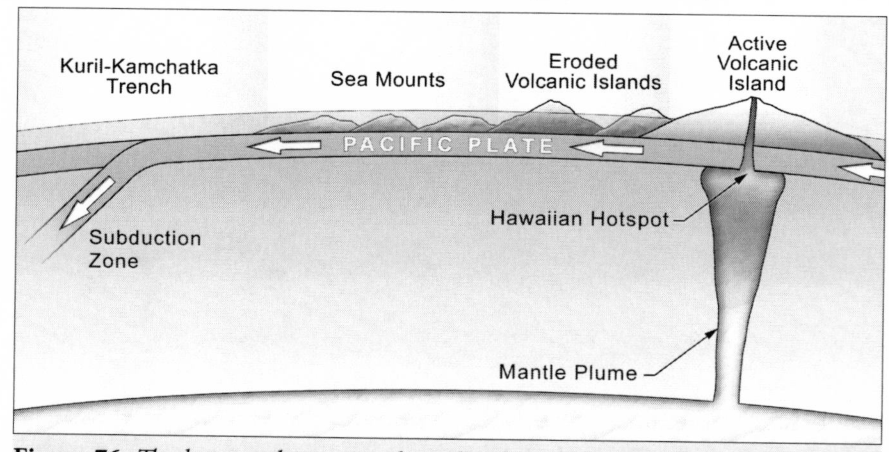

Figure 76. *The hotspot theory was formulated to explain undersea ridges like the Hawaiian–Emperor seamount chain, which truncate in the middle of a seafloor plate and exhibit volcanic activity. According to this theory, the Pacific Plate moves westward, where it subducts into trenches like the Kuril-Kamchatka Trench. As the Pacific Plate migrates west, it moves over an unseen subterranean hotspot that melts through it. Lava rising up through this ever-extending seam forms a chain of volcanic seamounts and islands.*

Plate traveled westward on its way to subduction zones in the form of seafloor trenches, the plate traveled across regions of super-heated magma often referred to as mantle plumes. This stream of super-heated magma melted entirely through the overriding plate like a torch cutting a path through sheet metal, allowing lava to flow upward through the plate (Fig. 76). As the plate continued to move across this hot spot, heading toward and beneath the Asian Plate, the lava flow left a linear scarred surface of seamounts and eroded volcanic islands in its wake, forming a hotspot ridge.

Cusp and Ridge Alignments Refute Subduction

Has the Pacific Plate actually migrated westward and subducted beneath the Asian continent? The first piece of evidence to the contrary is the Hawaiian-Emperor seamount chain's alignment with the cusp of a ductile fracture. According to the hotspot theory, the hotspot that formed the Hawaiian-Emperor Ridge would have to have randomly passed directly beneath this cusp. More precisely, the Hawaiian-Emperor Ridge is sliding beneath the Asian Plate along with the Pacific seafloor plate; if we were to move backward or forward through time, the ridge would intersect with the Asian coast at a point further to the north or south respectively. Granted, if this particular alignment were a lone instance there would be a possibility it was a random occurrence, but as was dem-

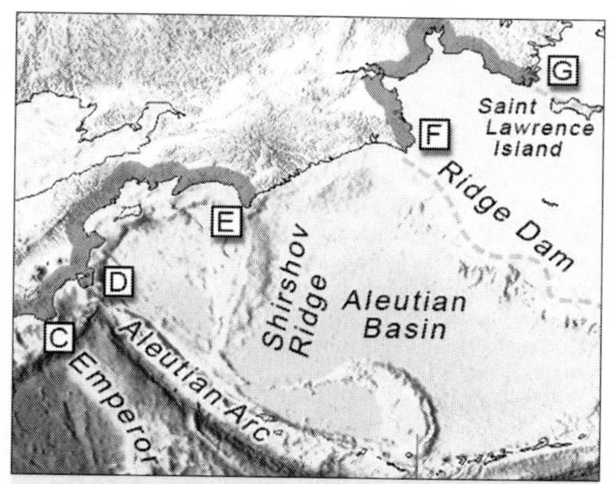

Figure 77. *Alignment of Seafloor ridges with ductile cusps in the Kamchatka Bering Sea region. According to plate tectonics, these alignments occur by chance. A few million years past or into the future and the Shirshov Ridge and Aleutian ridges would not have perfectly mirrored the bay spanning points D and E and in fact we would be fortunate to see one, let alone two cusp and ridge alignments.*

onstrated earlier, these alignments occur consistently throughout the region.

So as the Aleutian Basin and the Pacific Plate subduct beneath Asia, all ridges along the eastern coast of Kamchatka would randomly intersect the coastline migrating southward as the plates trend westward. But as can be seen in Figure 77, this would mean that five separate ridges would just happen to fall into perfect alignment with five consecutive coastal cusps at this point in time in Earth's existence. That is an extremely improbable coincidence, even more improbable considering how all the ridges that align nearly perpendicular to the coast align with cusps that extend out perpendicularly from the coast. The one exception in the midst of this array, the downward-angling Shirshov Ridge, finds itself aligned with the one cusp that extends downward precisely aligned with its center.

The Kamchatka region exhibits one of the highest concentrations of ductile cusping in the world, and it also exhibits the highest concentration of ridges intersecting a continental coastline. The simple fact is that in order for these features to consistently align in this region, there must be little to no plate subduction occurring along this portion of the Asian coast.

Seafloor Folding Versus Subduction

So how do we explain the extensive length of seafloor trenches that define the western edge of the Pacific seafloor plate? Until now, the belief that these trenches were formed where the Pacific Plate subducted beneath Asia has sufficed. But there is only one logical conclusion at this time that would allow for retained alignment and minimal shifting between the Pacific Plate and Asia. These deep creases in the seafloor can be explained either by minimal subduction or the radical notion that they are not truly subduction zones but are instead downward folds within the seafloor crust.

As we shall see, seafloor folding is the most reasonable means of reconciling the almost perfect alignment between ridge and ductile cusps across a trench while explaining the existence of other basic characteristics of seafloor trenches. This alternate dynamic will also move us toward a more unified theory explaining the creation of *all* seafloor ridges—unlike plate tectonics' suggestion that some ridges are plate boundaries while others are the result of unseen subterranean forces like hot spots.

The first notable characteristic of trenches, especially those that sit along the Western Pacific, is that they happen to exist at transitional zones between two seafloor plates lying on significantly different planes. For instance, west of the Kuril-Kamchatka Trench, the Okhotsk Basin sits at approximately 10,000 feet below sea level. Opposite the trench, the Pacific seafloor is nearly double that depth at approximately 20,000 feet. Why is this disparity in seafloor depths significant?

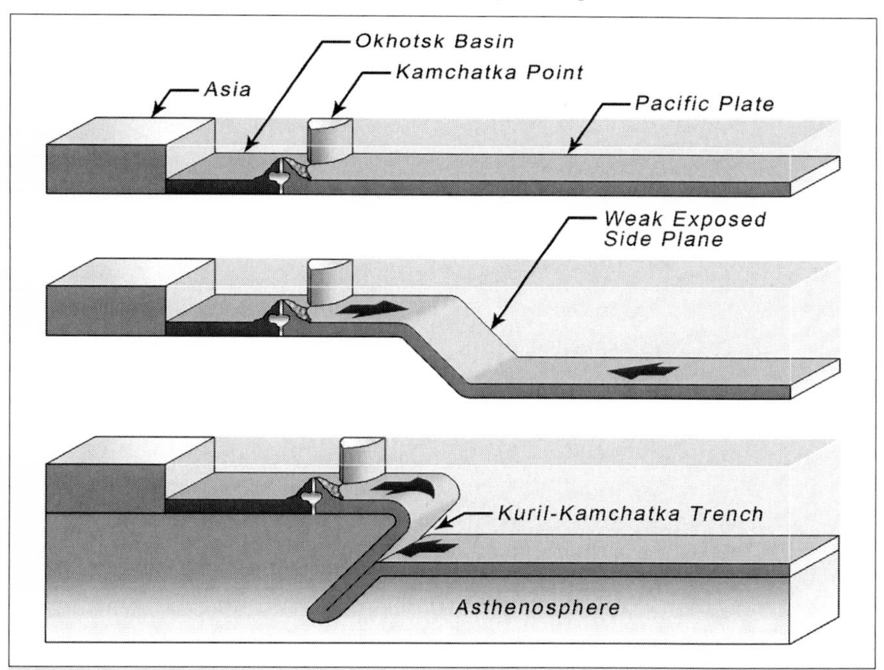

Figure 78. *Two plates on the same plane (top) are less prone to folding as they press each other against their rigid edges. When plates located on two separate planes converge (center and lower illustration), the rise that extends between the planes proves to be the vulnerable weak region, allowing the lower plate to fold in on itself along this rise as the upper plate pushes past and over the lower plate. Given the 10,000-foot disparity between the Pacific and Okhotsk seafloor, folding—not subduction—is the most likely genesis of trenches like Kuril-Kamchatka.*

Figure 78 provides the answer to this question by contrasting the effect of two converging plates that exist on the same plane versus two separate planes. In the top illustration, where we see both plates lying on the same plane, the two plates would resist converging since each plate would be pushing against each other where they prove most rigid. Each plate's mass aligns exactly alongside the other's, exposing no weakness.

On the other hand, if the plates exist on substantially different planes, as is the case with the Okhotsk and Pacific Plates, the weakness of one of the plates is exposed. The weakness in this case is along the lower Pacific Plate, where it slants upward and connects to Kamchatka. The lower two illustrations in Figure 78 demonstrate what occurs in this instance.

This simple principle can be replicated by placing a sheet of paper on two offset planes and pushing the sheet in on itself. Very simply, the upper plane slides and folds over the lower plane while the lower plane does the same beneath the upper plane. As applied to the Kuril-Kamchatka Trench, with the denser asthenosphere lying just below the seafloor plates limiting where the Pacific Plate can slide under the overriding plate, the Pacific Plate is folded tightly upon itself, compressed within the asthenosphere in a downward angle that extends beneath the overriding Okhotsk Plate. This phenomenon is depicted in the bottom illustration.

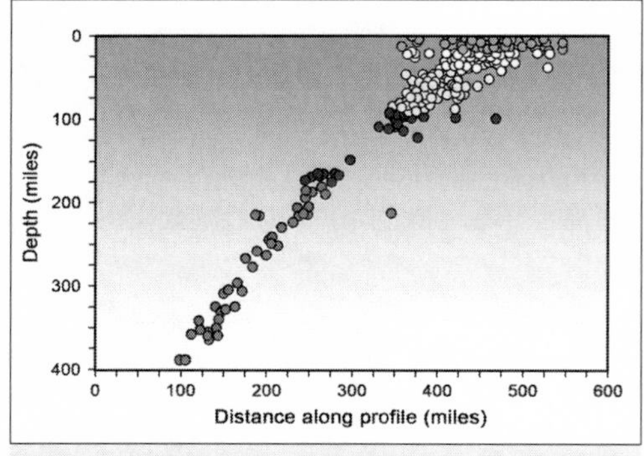

Figure 79. *Cross-section of recorded seismic activity extending well beneath the Kuril Islands believed to be evidence of plate subduction. This pattern of activity is commonly referred to as the Benioff zone and is believed to represent seismic activity occurring along or within the subducting Pacific slab.*

There are other phenomena occurring near trenches that scientists attribute to subduction, but I believe these are readily explained by seafloor folding. The first to discover these phenomena were seismologists Kiyoo Wadati and Hugo Benioff, and they have since become known as Benioff zones (Fig. 79). The seismologists found that there was increased seismic activity running parallel to seafloor trenches. In plotting the depth and location of these quakes, they realized a consistent pattern: There were multiple seismic events extending from the trenches downward beneath the overriding plate at a 40-60 degree incline relative to Earth's surface.

Operating under the assumption that seafloor trenches were locations where seafloor plates were subducting or sinking back into Earth's mantle, scientists postulated that the pattern of quakes was associated with seismic activity occurring along or within the subterranean slab as it slipped past the overriding plate and back into the mantle.

Operating under this new hypothesis that trenches are formed by seafloor folding, we can attribute these patterns of seismic activity to shifts within and around the fold of the Pacific Plate, which has a similar angled descent extending beneath the overriding Okhotsk Plate.

The second trench-related phenomenon currently attributed to seafloor subduction is island arc formation. Like the Benioff zone, this feature runs parallel to seafloor trenches, but on the overlying plate. Figure 80 distinguishes the old subduction theory from the new seafloor fold theory by the placement of the plate boundaries. The plate tectonics subduction theory places the boundary at the trench where the Pacific Plate is believed to slide past the Okhotsk Plate (Fig. 80 left).

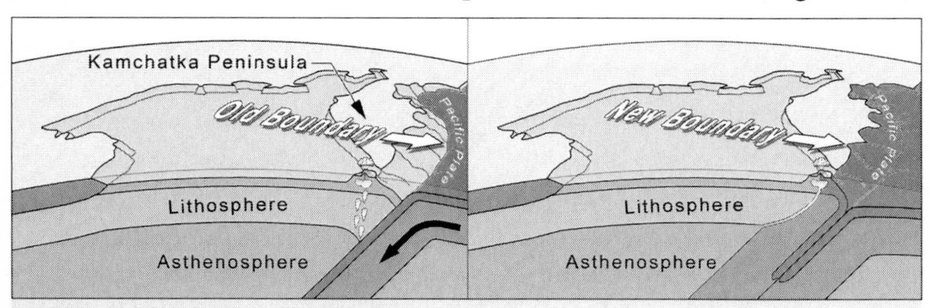

Figure 80. *Old (left) and New (right).*
Old: *Plate boundary at trench.*
New: *Plate boundary above trench forming Kuril-Kamchatka Island Arc.*
Old: *Benioff zone related to friction from slab subduction.*
New: *Benioff zone related to shifting within Pacific seafloor fold.*
Old: *Kuril Island Arc formed by friction melting through upper plate.*
New: *Kuril Island Arc formed by magmatic seepage in plate boundary.*

In this scenario, scientists theorize that friction from the subducting plate generates super-heated magma and, similar to the hotspot theory, this material cuts through the overriding plate to form a ridge or chain of islands called an island arc.

The seafloor folding theory places the boundary along the eastern coast of Kamchatka, where the Pacific Plate is firmly attached, and finds it extending out along the island arc above the trench. So instead of a ridge creation theory that requires two completely different unproven sources for magma melting through crustal plates—slab friction and mantle plumes—our new theory offers a far less complex uniform theory of ridge formation. It maintains that island arcs and "hotspot ridges" are simply plate boundaries that, due to stress between the two plates, have allowed magma to seep through to form islands and raised ridges.

With all the pieces now in place, let us take another look at the combined dynamics that contribute to Kamchatka's separation from the Asian continent. As shown earlier, we propose that Kamchatka remained attached to the Pacific Plate along its eastern coast as the Asian continent began to pull away. Figure 81 demonstrates how relatively basic this newfound relationship between continental and seafloor crust transformed the region. Once again, this is a relationship similar to that of a wooden picture frame with canvas attached; in this model, Kamchatka corresponds to a piece of the frame remaining attached to the canvas after the frame has been broken.

Figure 75 shows a rip extending out perpendicularly into the seafloor crust when the continental crust fractures parallel to the seafloor in an opening-mode fracture. Conversely, Kamchatka represents an in-plane shear where a portion of the continental plane recedes away from the seafloor while another portion remains. In this scenario, the rigid seafloor crust is unable to remain attached to the receding Asian continent.

Fracturing the continental crust in this fashion is similar to the depicted break in an artist's frame (Fig. 81). The bond between frame and canvas rips away downward along the receding frame, forming an open wedge or "V" extending between the exposed edges of the frame and canvas. We see exactly this same open wedge between Sakhalin—

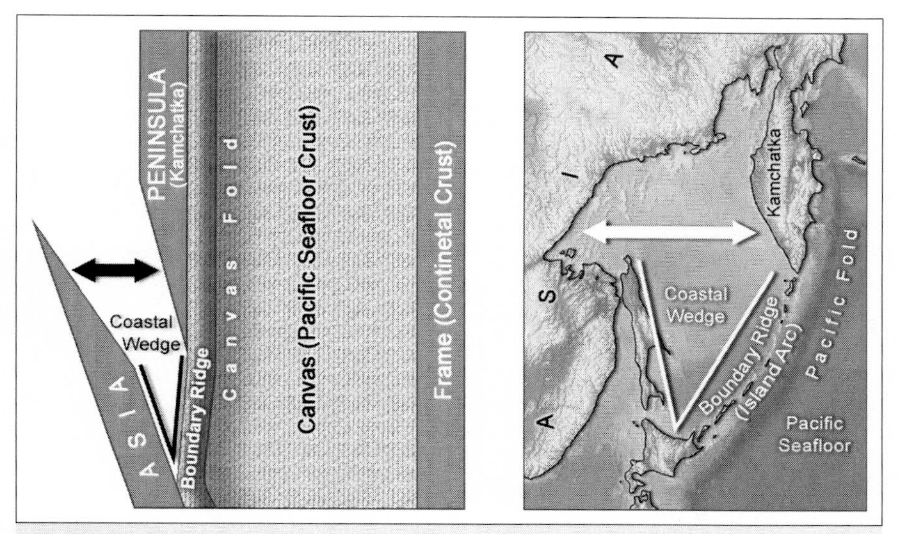

Figure 81. *Like canvas to a wooden frame, seafloor crust is attached to continental crust. If the frame or crust fractures away from the canvas or seafloor, the remaining frame attached to the canvas or seafloor forms a peninsular splinter. Where the continental crust pulls away and breaks free from the Pacific crust, a "V" forms, extending out beyond the peninsula. Unlike the artist's frame, magma below the continental crust is exposed, hardening and filling the void with new seafloor crust. In both instances, a boundary ridge extends directly downward from the point of the peninsular structure to where the canvas/seafloor meets back with the frame/continent (Northern Japan, originally merged with the Asian mainland).*
Another shared attribute is demonstrated as the canvas is pushed beneath the broken frame: The fold or trench runs parallel below the peninsula, the boundary ridge, and the rest of the frame.

which was at one time the coast of Asia before it broke free of the continent—and the Kuril Island Arc.

Unlike the wooden frame that exposes a void between the frame and severed canvas, the receding continental plate exposed the mantle's molten magma lying beneath. The magma subsequently filled in the void and cooled to create new seafloor crust in the form of the Okhotsk Basin. Similar to the edge of the exposed canvas, the Kuril Island Arc marks the edge of the Pacific Plate, the boundary between it and the newly formed Okhotsk Plate. The ongoing shifting between these two plates along their shared boundary has allowed magma to seep up between them forming the Kuril-Kamchatka arc.

The frame-and-canvas model in Figure 81 also incorporates a fold, illustrating the dynamic of seafloor folding forming an adjacent trench where the canvas has slid beneath the fractured side of the frame. The frame-and-canvas model can prove extremely useful in understanding the true genesis of seafloor features such as ridges and trenches. Similar to the canvas, the seafloor is attached to the continents; movement or fracturing of the continental crust generates all seafloor expansion, with ridges signifying either boundaries between seafloor plates or the central expansion ridge within an expansion zone.

Hawaiian-Emperor Boundary Ridge

At first, this logic seems solid. However, when we look at the Hawaiian-Emperor seamount chain (Fig. 82) it is difficult to reconcile its existence with this new theory. We find ourselves revisiting the same dilemma geologists encountered when they originally attempted to reconcile the ridge with a belief that all seafloor ridges were plate boundaries.

With the introduction of a new dynamic providing a link between ductile fractures and boundary ridge formations, we have an opportunity to approach the problem from a far different perspective. Most boundary ridges should run together, forming a wedge, as demonstrated earlier in Figure 75. We find this closed wedge in the region lying between the Shirshov Ridge and Aleutian Island Arc (Fig. 77). The deformity of the wedge shape implies shifts in movement between the two continental plates as the Bering Plate was forming, but nevertheless, we should expect a ridge to be closed in this way if it is truly a boundary ridge.

Of course, there are instances where the boundary exists but no ridge is visible. We should not expect boundary ridges to uniformly allow magma to seep through and above the seafloor. This is clear in looking at the Shirshov Ridge, where we can see that wide swaths of magmatic material appear at each of its ends, but the ridge thins and completely disappears in the center. Based on the shape of the Okhotsk Plate, the center point of the Shirshov Ridge would appear to be an

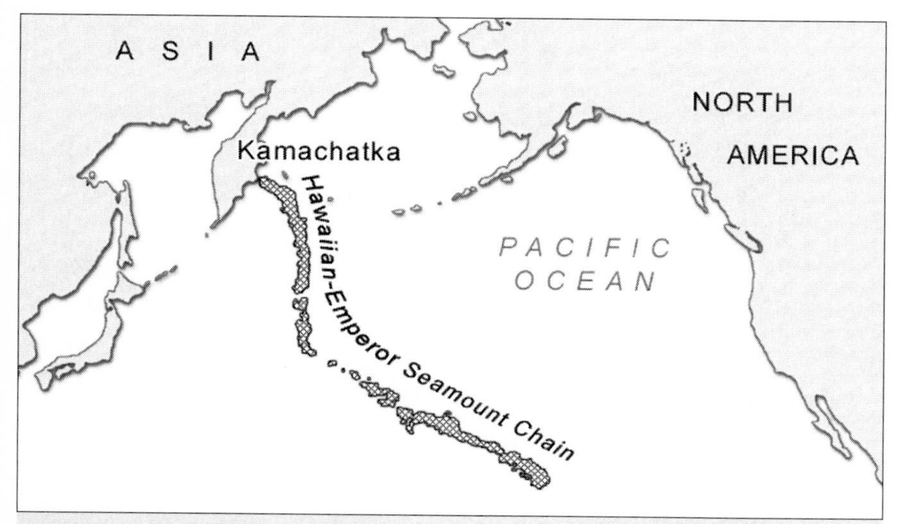

Figure 82. *The isolated Hawaiian-Emperor seamount chain extending out to the center of the Pacific has, to date, been explained as a hotspot ridge. It is, however, possible to fit the feature within the new unified theory and identify it as a boundary ridge.*

active pivot point during plate movement. Pivoting of the plate from side to side would maintain a tight fit at the pivot point, or fulcrum, allowing little to no magmatic seepage. Meanwhile to either side, the boundary ridge would alternately open and close, allowing large flows of magma to seep upward and creating the existing side-to-side ridge build up.

This dynamic may explain similar thinning in the Hawaiian-Emperor seamount chain at a similar fulcrum point in its center, but it does not explain the abrupt truncation of the chain at the Hawaiian Islands. There should still exist the other half of the ridge—a mirrored sister ridge of similar length framing an expansion wedge, as the model in Figure 75 demonstrates. The northwesterly rise of the Hawaiian-Emperor seamount chain back to the Asian continent should be matched or mirrored by a similar ridge to the east that similarly extends back to the north, bracketing the Pacific seafloor east of the Hawaiian-Emperor seamount chain.

Aiding the search is a seafloor rife with well-defined lateral fracture zones. As demonstrated earlier in Figures 65 and 66 (pp. 145, 146), we should be able to track directly along these fracture zones to find its mirroring ridge. This directs our attention to a unique ridge in the Pacific that similarly extends and then abruptly truncates south of the Molokai Fracture Zone (MoFZ): the Baja Peninsula. Identifying the Baja Peninsula as the eastern sister ridge is significant, because it requires that the seafloor between the two ridges exists as an expansion zone with an obscure central divergent boundary, which runs counter to our current understanding.

However, the fact is that we have two substantial truncated ridges extending southward into the Pacific. This alone is intriguing, but they also in like manner extend just below the MoFZ, increasing the possibility that the two ends were once merged together at a central divergent boundary in the North Pacific. We can recreate that original merged state by reducing the distance between the twin ridges along the shared fracture zones. Figure 83 demonstrates that when we do so,

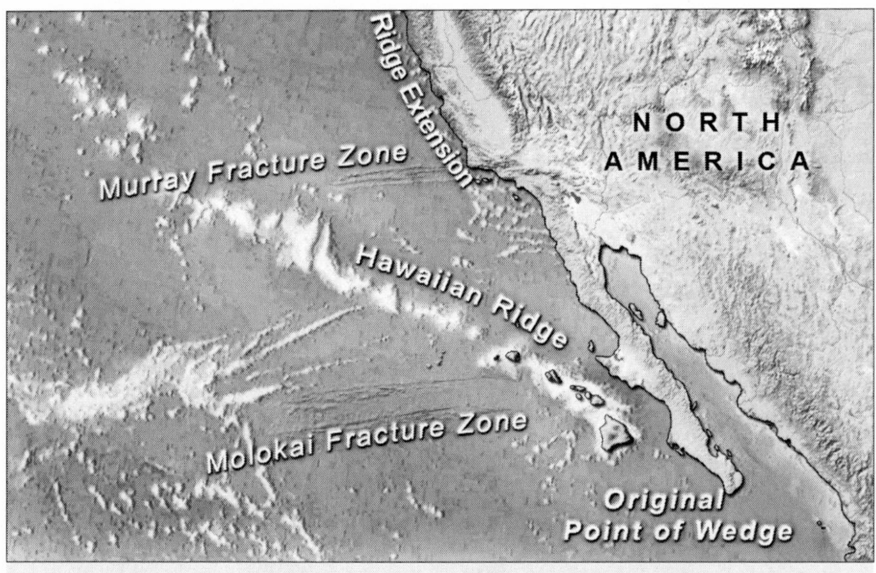

Figure 83. *Two ridge segments, one from the Hawaiian Ridge and the other the Baja Peninsula, extending from the Murray Fracture Zone to just beyond the Molokai Fracture Zone. These were likely once merged together to form the lower boundary of an expansion wedge.*

the Hawaiian Ridge comes back to rest in what appears to be a natural fit—pocketed into the side of the Baja Peninsula, thus reconstructing the original closed point of the Northern Pacific expansion wedge.

For this to be possible, of course, we have to assume the existence of a central divergent boundary between these two features. It should be noted that every ocean in the world has a visible expansion ridge lying between opposing continental plates, but the North Pacific, one of the largest bodies of water on the planet, exhibits no such ridge. Physical obscurity, however, should not negate the existence of a divergent boundary. In the South Pacific, the divergent boundary is very unimposing and nearly imperceptible in places. The same is true in the Southern Ocean between Australia and Antarctica.

Accepting the possibility that a divergent boundary is not necessarily defined by a perceptible ridge, we should still expect to see sufficient signs of mirroring on each side of the expansion zone verifying its existence. Regarding the Hawaiian-Emperor seamount chain, we should expect that an expansion zone extends its full length, thus mirroring of the seamount chain should extend much further to the north than just the length of the Baja Peninsula. Topographic maps actually do suggest that the Baja Peninsula is merely the southernmost portion of a very extensive ridge lining the western coast of North America (Fig. 84).

At the northern end of the Baja Peninsula at the Murray Fracture Zone (MuFZ), there is a shift in the ridge and a sharp drop in elevation and detail in the southern portion of the Pacific Coast ranges. A clear transition similarly occurs along the Hawaiian-Emperor seamount chain at the MuFZ, where the ridge thins substantially and may demonstrate an alternate genesis than the earlier discussed fulcrum effect. The lower elevation may be due to reduced tectonic activity when these two portions of the ridge were still merged and just coming into existence. This northern extension of the Baja Peninsula does not end here; it extends all the way to the Coast Mountains of Canada and Alaska, where it finally folds over, creating the Kodiak Islands in the far north.

Significantly, this distinct range saddled up along the coast of North America is roughly 3,900 miles in length. Measuring the Hawaiian-Emperor seamount chain from the edge of Kamchatka to Hawaii also

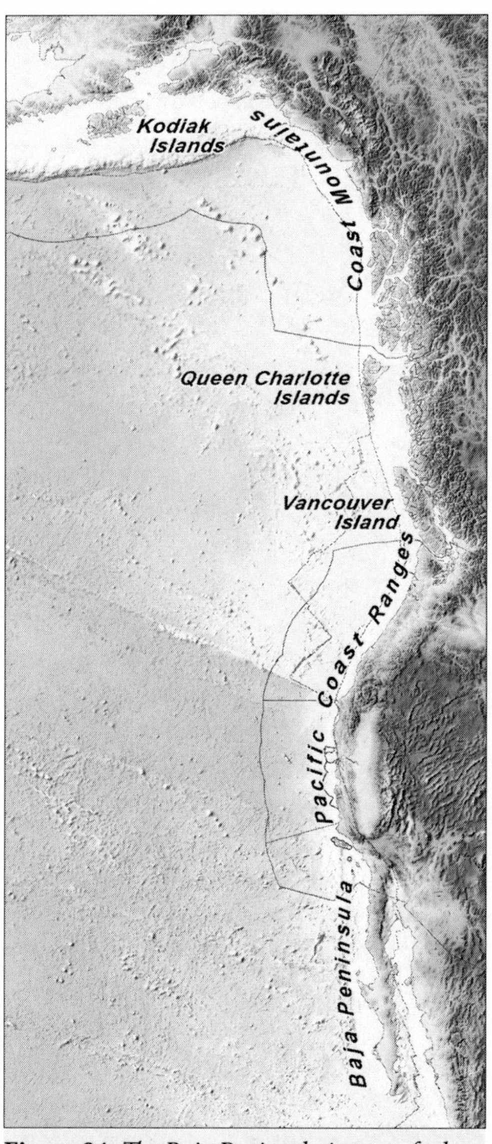

Figure 84. *The Baja Peninsula is part of a long narrow coastal range that extends northward to the Pacific Coast Ranges, Vancouver Island, Queen Charlotte Islands, Canada's Coast Mountains, and Kodiak Island forming the North American Coastal Ridge (NACR).*

returns a measurement of roughly 3,900 miles, making the North American coastal ranges and Baja Peninsula a highly viable candidate for a sister ridge. The fact that it lies above sea level is a bit of a concern, but it is not unique. The answer to why this sister ridge lies almost completely above sea level hugging the coast of North America will become clear later in this work and relates to a unique aspect of the Pacific Ocean: the western seafloor sits much deeper. The seafloor adjacent to the eastern sister ridge, which we will refer to as the North American Coastal Ridge (NACR), sits at between 9,000 and 14,000 feet below sea level, while the seafloor adjacent to the east of the Hawaiian-Emperor seamount chain lies 15,000 to 19,000 feet below sea level.

Meanwhile, there is additional evidence that the NACR is truly the sister ridge we had set out to find—evidence that will also aid in locating and identifying the existence of a previously unknown central divergent boundary in the North

Pacific. It exists in the form of additional distinct mirrored features lying between the two ridges that similarly index off shared lateral fracture zones.

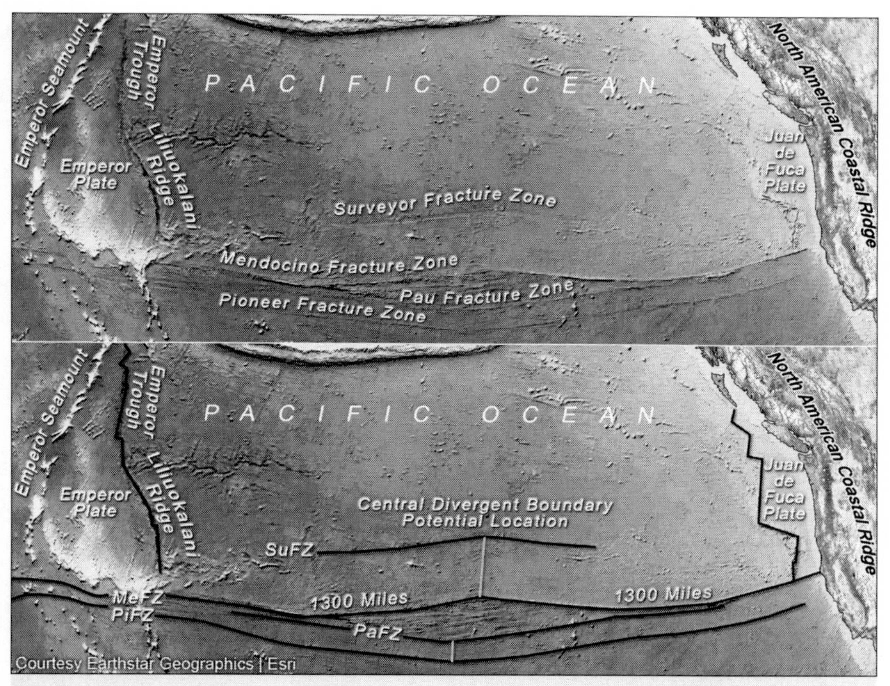

Figure 85. *Substantial geographic mirroring upward from the Pioneer FZ, centered between the Emperor seamount and the North American Coastal Ridge. It reveals that the Juan de Fuca Plate has a mirrored plate in the west that also extends up from the Mendocino FZ: the Emperor Plate. The two plates are similarly sandwiched between minor inner ridges tapering upward and outward to the outer major ridges.*

It is in the Northeast Pacific that we find the most extraordinary examples (Fig. 85). First, we have mirroring within the fracture zones themselves. In this truly unique instance, we find a lateral diamond-shaped region sandwiched between two primary fracture zones. The Mendocino and Pau Fracture Zones (MeFZ, PaFZ) form the diamond's northern and southern borders respectively, and their two central points mark a centerline from which the fracture zones mirror themselves east to west. Paralleling these fracture zones are two additional fracture zones. The Surveyor Fracture Zone (SuFZ) lies 240 miles north of the MeFZ, while the Pioneer Fracture Zone parallels the PaFZ at between 70 to 90 miles to the south. All four fracture zones mirror east to west of their central points, signifying the possible location of a central divergent boundary.

Even more intriguing, lying roughly 1,300 miles to the east and west of this potential expansion zone along the MeFZ, we find two separate highly detailed ridge and trough formations rising up and away from this central divergent boundary. Both ridges similarly extend upward from the MeFZ with no apparent bleed over to the south. The ridge lying at the eastern extremity, near the coast of North America, differs from the one lying in the west; greater shifting has occurred along lateral transform faults, creating a far more jagged structure. In contrast, only minor shifts are discernible in the Emperor Trough and Liliuokalani Ridge lying to the west.

The plate defined by the minor ridge in the east and the NACR is a well-studied stretch of seafloor known as the Juan de Fuca Plate or microplate. Within the plate tectonics model, the relatively small 240-mile-wide Juan de Fuca Plate is thought to be a unique feature. It is the remaining portion of an immense hypothetical Farallon Plate, over 1,000 miles wide and perhaps twice that height, which is thought to have subducted beneath the North American Plate.

There is a similar mirrored microplate, the Emperor Plate, nestled between the Emperor seamount and the inner Emperor Trough-Liliuokalani Ridge. Its existence suggests that the Juan de Fuca Plate is mirrored by a sister plate—meaning that the Juan de Fuca Plate was never larger than it is today or part of any massive subducted plate. Much like its sister plate, it remains intact, nestled above the Mendocino Fracture Zone alongside a major ridge formation.

While the mirroring in the north is fairly clear, more evidence of a North Pacific divergent boundary lies to the south between the Murray and Molokai Fracture Zones (Fig. 86). Three linear seamounts extend westward a short distance off the MuFZ before plunging to the southwest. The two outer lines of seamounts are much shorter than the central seamounts, indicating that they may be mirrored features. Meanwhile, the central grouping exhibits a much larger array of

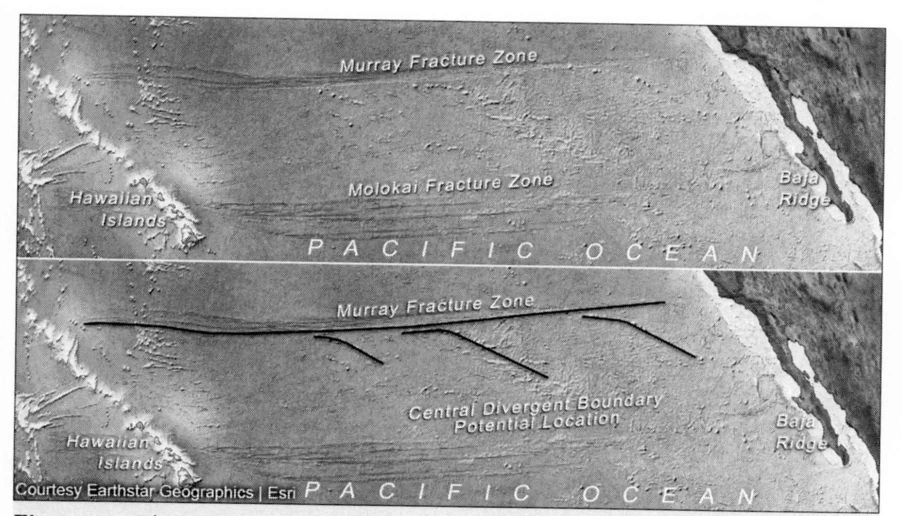

Figure 86. *Three mirrored ridges extending down off the Murray FZ with signs of increased seismic activity surrounding the central ridge, which suggests that it is a portion of the elusive central divergent boundary lying in the Northern Pacific.*

seamounts, signifying increased tectonic activity, which may suggest a divergent boundary. This may be the only portion of the divergent boundary that is visible, at least at current mapping resolutions.

Beyond these examples, it is difficult to determine the entire layout of the North Pacific divergent boundary using current maps. However, the existence of the North American Coastal Ridge mirroring the Hawaiian-Emperor seamount chain in the North Pacific, along with the mirrored Juan de Fuca and Emperor microplates that nestle in along these ridges, presents a very strong case for the boundary's existence. A divergent boundary in the North Pacific would create a global consistency where all continental plates are moving apart. In other words, all seafloor crust would demonstrate central expansion.

This consistent global central-seafloor expansion in tandem with the cusp and ridge alignments across trenches in the Kamchatka region provide clear proof that plate subduction has not occurred at any significant level that would support the plate tectonics model. This observation potentially marks the end of plate tectonics and leaves open the door to the lone remaining Earth dynamic: Earth expansion.

Expanding Earth

As mentioned earlier, the discovery of seafloor expansion brought about two viable theories that accounted for ancient seafloor crust displaced by the generation of new seafloor crust. The first theory was based on the premise that the earth remained static in its overall spherical dimension, which left only two locations where the old crust could come to rest.

The first location was on the earth's surface but was easily rejected, as the planet does not exhibit adequate accretion of crustal material on its surface. For instance, the expansion and formation of the Atlantic seafloor alone would have required the displacement of a similar amount of seafloor elsewhere. If it had remained on the surface and folded in upon itself—as one would expect based on the current view of mountain creation, or orogeny—then the crustal material should still be present in the form of mountain ranges dwarfing the Himalayas in both size and number. The absence of these crustal accretions necessitated the second location. The old seafloor crust had to return from whence it came: back into Earth's mantle. With the discovery of deep seafloor trenches, scientists appeared to have confirmation of seafloor subduction, and plate tectonics was on its way to scientific consensus.

The other viable theory that was pushed aside to make way for plate tectonics was the expanding Earth theory. Now with evidence introduced which demonstrates the lack of significant plate subduction, the expanding Earth theory becomes the lone viable theory. Absent plate subduction, we now have to accept that all newly generated seafloor crust remains on Earth's surface. Since we do not see signs of excessive orogeny, planet Earth must have increased in size to accommodate both ancient seafloor and all new seafloor emanating from the ocean's divergent boundaries.

The expanding Earth theory maintains that Earth originally existed at roughly half its current size. It assumes that all of Earth's continental crust existed as a single unified plate, much like that of other planets in our solar system. Based on Earth's 58 million square miles of land surface, this would project to a sphere of about 4,300 miles in diameter, 54% Earth's current 7,918-mile diameter.

This theory carries with it major challenges, which have existed from the time of its initial proposal. For instance, how do we account for a planet growing in size when we have no evidence of planets experiencing expansion? While space appears to be expanding as a result of the Big Bang, planets, stars, and galaxies are believed to remain the same size. We would also have to identify the catalyst of the expansion process. Even more importantly, scientists have taken high-precision measurements of the earth and confirmed that it is not getting larger. Therefore, once the planet began the expansion process—which would see an enormous surge of deep subterranean energy emanating from Earth's core—what possible mechanism could terminate this chain reaction once it reached its current static size and prevent it from continuing on toward a volatile end?

We will tackle these complex issues shortly, but before we do, we are going to perform a global examination of Earth's geography building on the geological and geographical observations described in this chapter. We will witness substantial evidence of a consistent pattern of continental plate separation—and a lack of evidence supporting plate convergence, a staple of plate tectonics.

CHAPTER 11

EARTH DECONSTRUCTED

In this chapter, we will continue building a case for Earth expansion. Our goal will be to piece the planet back to its original state, as a single unified shell, using the tools acquired from the previous chapter. Considering all the initial revelations brought to bear via Kamchatka's formation, it is logical to start by searching for consistency in peninsula formations throughout the globe.

Peninsular Dynamics

To the south of Kamchatka lies the Korean Peninsula. Unlike the Kamchatka Peninsula, geologists do not propose it erupted up from the seafloor. They shy away from a unified theory of peninsular genesis by proposing a more complicated process that embraces the chaotic to-and-fro plate movement envisioned by proponents of plate tectonics.

According to a 2003 research paper entitled *Early Ordovician Paleogeography of the Korean Peninsula,* co-written by Sang Min Lee and Duck K. Choi, the majority of the peninsula originally existed in the form of two fragments lying thousands of miles apart. This came about because the study, based on fossilized faunas, determined that two mountain groupings, or massifs, on the peninsula were populated with two very distinct collections of marine organisms known to inhabit

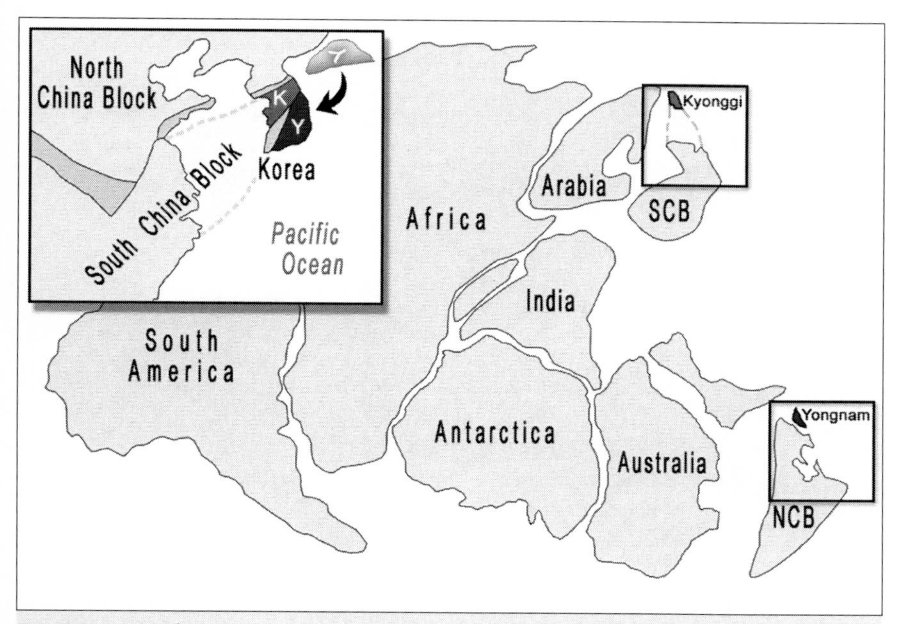

Figure 87. *A plate tectonic view of the Korean Peninsula formation (*Early Ordovician Paleogeography of the Korean Peninsula - *Sang Min Lee and Duck K. Choi). Keeping with the view that continental plates are continuously meandering around Earth's surface in chaotic fashion, this theory maintains that the North and South China Blocks (NCB/SCB) each drifted along with one small fragment of the Korean Peninsula. When the two blocks of China finally came together, the Kyonggi fragment (K) associated with SCB happened upon an alignment with an extension of NCB that had a similar width. Shortly after this event, the Yongnam fragment (Y) associated with the NCB miraculously rotated over to cap off the peninsula, creating its final form (see inset).*

entirely different environs. The Kyonggi massif (Fig. 87) seemed to have once been populated by deep sea organisms also found in a large region of China referred to as the South China Block (SCB). Meanwhile, the Yongnam massif contained the fossils of shallow, water-dwelling organisms that once inhabited a separate part of China known as the North China Block (NCB).

The existence of two distinct environments suggested that the SCB and the Kyonggi massif were originally paired together in a remote region, sharing an environment vastly different from the one shared by the similarly paired NCB and Yongnam massif. At some point in

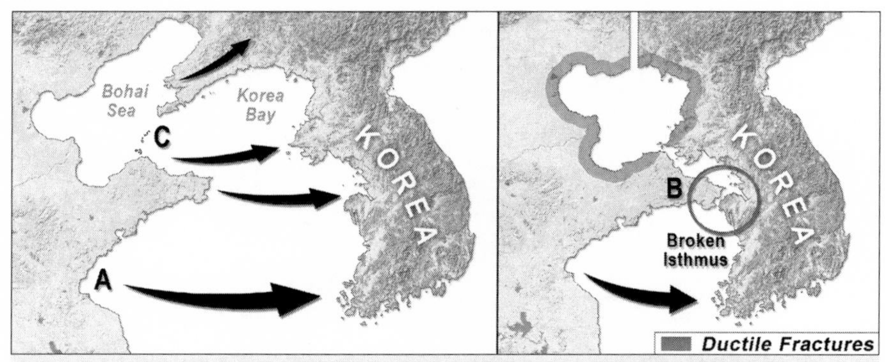

Figure 88. *Like Kamchatka, Korea is a peninsula rotating out and away, counter-clockwise, from a conforming Asian coastline. The southwestern portion of the peninsula conforms too well in size and shape to an angled pocket in the coast of Asia (A) to be coincidence. Similar distance up from each point lie two very similarly sized tabs of land that once formed an isthmus (B encircled). The remaining open areas, the Bohai and Korea Bay, are ductile fractures. A set of islands form a cusp leading toward a sister cusp in the north (C), revealing that the Bohai Sea was once a closed inland sea or ductile void ripped open. Likewise, the arced shape of the Korea Bay reveals that it too is a ductile fracture. The visual evidence suggests that Korea began its existence with its tip plucked from the Asian continent. Further stress saw the opening of an inland Bohai Sea. Korea then began to separate completely, with a reluctant isthmus giving way. As Korea continued to be pulled free, the Korea Bay tore open, creating the region's final form.*

the ancient past, the two paired landmasses set off toward each other on a collision course. The theory proposes that when the SCB finally did collide with the NCB, the Kyonggi massif simultaneously aligned and merged with an outcropping of land exhibiting similar width. Next the Yongnam massif suddenly separated itself from the NCB and began pivoting around the recently seated Kyonggi massif, finally finding its way to the end of the stack to cap off the Korean Peninsula (Figure 87 Inset).

It is a theory strongly based on assumptions from ground level observation, but once again, it appears that the clearer picture may exist viewing the planet from above through the use of maps. Like Kamchatka, the evidence suggests that the Korean Peninsula is also a singular fragment that has partially fractured or splintered out from the Asian continent.

Looking at the peninsula's southwestern corner, it is obvious that it pivots perfectly into an accommodating pocket cut into the Asian coastline (Fig. 88 A). Halfway up the peninsula's western coast is a notch whose southern end is composed of a 32-mile wide tab of land partially separated from the peninsula. Similar to the broken isthmus on Kamchatka, this feature conforms to a similarly sized squared-off tab of land lying directly across from it (Fig. 88 B), suggesting that the coastal feature had at one time been nested into the side of the peninsula.

While brittle fracturing is responsible for forming this lower region of the peninsula, ductile fracturing accounts for all remaining coastline lying to the north and further west. Korea Bay is a standard arced ductile fracture that would have been the last development in the peninsula's formation, torn into the coast as the peninsula continued to pull free from the mainland in a counterclockwise direction.

Meanwhile, the Bohai Sea is a series of ductile fractures, as evidenced by the two arced bays in the south; the cusping at the mouth of the bay is the most significant ductile feature (Fig. 88 C). A series of islands extends from the southern shore toward a narrow piece of land in the north, thinning out as they go. This clearly indicates that this entire sea existed for some time as an isolated body of water or depression, a ductile void, before tearing completely open and merging with Korea Bay.

Positioned in between Kamchatka and Korea lie the islands of Sakhalin, Hokkaido, and Japan. Though not connected, these three islands separated from the Asian coast in a similar fashion. They most likely existed in the form of a singular peninsula for a time, with Sakhalin in the north anchoring the landmass to Asia. Like Kamchatka and Korea, forces have separated the landmass from Asia, swinging it out in the same counterclockwise direction. These same forces also eventually led the peninsula to fragment into its current form.

Geologists do believe that Japan separated from the Asian mainland, but they incorrectly conclude that the base of the island (D) rested in Korea's northern bend (C). The base actually sat 280 miles further down the coast. Keep in mind that Japan could easily have conformed to the Asian coast, because the bend at C did not exist until after Korea began to separate from the mainland.

We can confirm Japan's original location because within the span lying between points A and D in Figure 89 lie traceable links in the form of isthmi both broken and plucked from the opposing landmasses. With the two previous peninsulas, we were able to demonstrate that broken isthmi can be a reliable tool in tracing ancient connections between landmasses and there are two such instances here.

Figure 89. *The islands of Sakhalin, Hokkaido, and Japan once formed a singular peninsula that, like Kamchatka and Korea, has separated in counterclockwise fashion from Asia. Points A through D were shared points, though distortions like an outward-protruding Korea no longer allow complete coastal conformance.*

The first can be seen at the bottom end of the shaded span (D) in Figure 89, where a small upward point of land extending from Pohang, South Korea once likely linked to Hirado Island at the southern tip of Japan. Figure 90 shows an enlarged view of the other broken isthmi. Starting at point A, we find another example of the outstretched peninsulas that once formed an isthmus.

Beyond this, we find two instances of a new isthmian dynamic. Instead of two outstretched coastal points, as seen in a broken isthmus, we find squared-off stubs or tabs of land that align with squarish pockets on the opposing landform. They appear to be interlocking tab and notch sets. Point B locates the tab on the Asian mainland, while C finds the tab extending from the Japanese coast. The fact that the linear distances between all four of these isthmian features (A, B, C, and D) on the Asian continent are relatively the same as those of the corresponding features on Japan's western coast confirms that these were once interlocked.

Figure 91 illustrates six of Earth's largest instances of the isthmian tab phenomenon. Although the additional four examples do not have notches lying directly across from them, if we have properly identified the cause of these structures, locating notches of similar size may help determine the coastlines that were originally merged with these tabs.

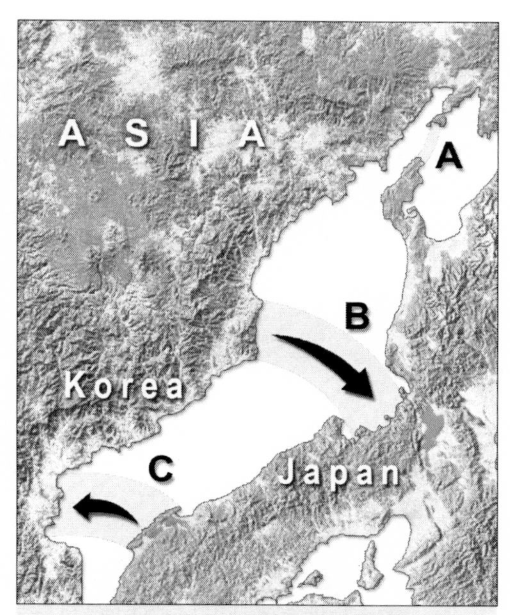

Figure 90. *A broken isthmus in the north (A) and a pair of apparent interlocking tabs and notches in the south (B, C) reveal Japan's original fit along the Asian coast. The distance between A and B is the same on both coasts. The same holds true for distances between B and C as well as C and D (ref. Figure 89).*

The features are clearly isthmi in various stages of formation that have plucked free a portion of the opposing coast during separation. The examples in Gwadar and Ormara, Pakistan (Fig. 91 bottom right) appear to have stretched nearly to the breaking point before the opposing coastline gave way. These two examples provide the best confirmation that we are looking at ductile stretching of an isthmus: The arcs along each side of these coastal points form what is known as necking, a clear indication of ductile stretching. These arcs lying below the mushroom cap are also prevalent on the remaining four tabs, though not stretched nearly to the same extent (Figure 91 A-A). The other indication of ductile stretching is the consistent presence of a low-lying swath of land between the arcs. This neck area (Fig. 91 B) is stretched to such an extent that the thinness of the continental crust leaves the region lying nearly flat, often just a few feet above sea level. A clear sign of a potential ductile void formation is also visible within the ductile swath of the Shimane Prefecture, where more extensive thinning of the crust has created depressions filled by freshwater lakes (Fig. 91 top left).

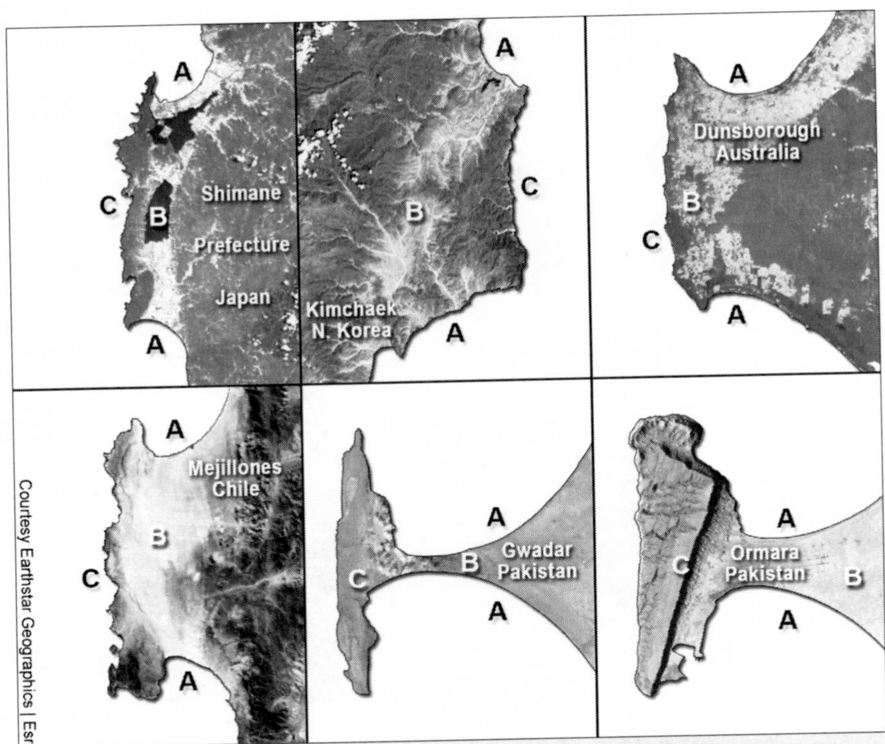

Figure 91. *The two top left images are ductile tabs associated with the Japan/mainland separation in Figure 90 followed by four others located around the globe. These tabs of land exhibit clear signs of ductile stretching in the form of varying degrees of necking or arcing along the extended sides (A) with the land between stretched to such an extent that it forms a narrower neck (B). This neck of land is relatively flat and lies close to sea level.*
Unlike the two isthmian central fractures observed on Kamchatka, these isthmi were stretched until a brittle fracture on the opposing continental mass occurred allowing a fragment, normally part of a mountain formation (C), to be plucked free. This normally leaves these tabs mushroom shaped with the extracted range wider than the isthmian extension.

Perhaps the most intriguing aspect of these stubs is the end composed of highland ranges (Fig. 91 C). It would seem from these six instances that the weakened crust found in the valleys between mountain ranges and linear hill formations is where both ductile thinning and fracturing occur.

Figure 92 provides an oblique view of the tab extending from the

Shimane Prefecture, Japan. Here we can see that the western coast of Japan is lined with a range of foothills, while opposite a light-shaded ductile zone lies a short range of hills on the end of the ductile tab. The ductile zone originally existed as the downward fold or valley between these hill formations.

Figure 92. *Shimane Prefecture. Foothills to the right of the ductile zone align with the foothill-lined Japanese coast north and south of the isthmian tab. Had the downward fold of land been more brittle, the fracture would have broken cleanly, forming an uninterrupted foothill-lined Japanese coastline. Instead, this downward fold is a ductile zone that extends out to a rugged fragment of the Asian continent.*

As in other instances of isthmian formations, had significant fracturing previously existed in this valley region, Japan would have had a clean break at the base of these foothills, and the coastline would remain uninterrupted by this protrusion of land. Yet instead of these ductile valley zones stretching out to a central breaking point, the central portion of the isthmus remains intact, with hills or upward folds that once rested at the opposite side of the ancient valley plucked from the opposing landform at what would likewise be another downward fold.

Figure 93 demonstrates this basic dynamic and the two eventualities of isthmian formations. In both instances, large-scale brittle fracturing occurs as two landforms begin to separate. This fracturing typically occurs along valley basins where folding has weakened the continental crust. In illustrations 1 through 3, we see the initial separation, with a portion of the valley resisting due to few or no preexisting fractures. Illustration 4a demonstrates the central fracturing observed as the ductile material stretches to its breaking point. Meanwhile, illustration 4b demonstrates the alternative outcome: crust weakened by a downward fold on the opposing landform fractures, allowing the plucking-out of a segment of mountainous region as wide as or wider than the attached ductile isthmian material.

This demonstrates two dynamics linked to ductile isthmi: the isthmian ductile fracture and the isthmian extraction. Understanding

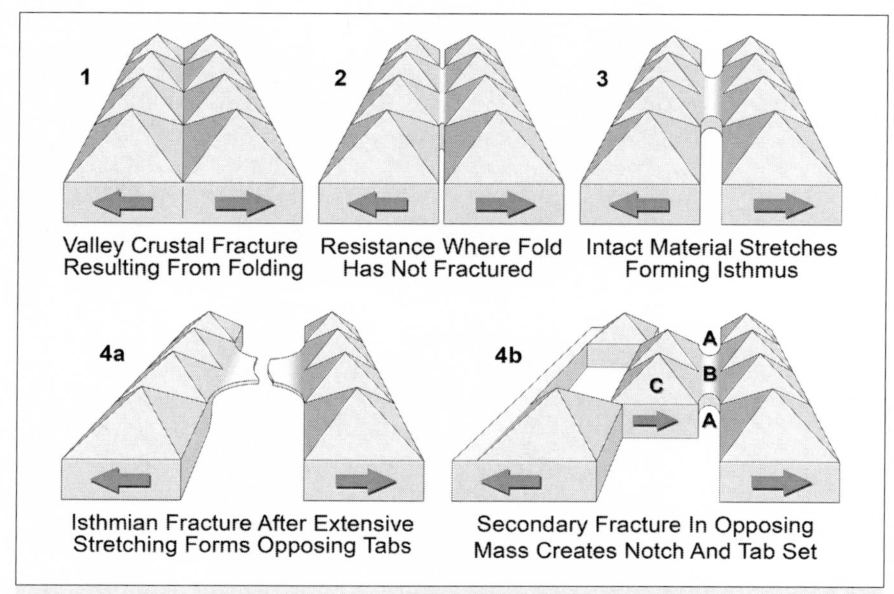

Figure 93. *Partial fractures can occur along a linear downward fold or deep valley lying between two mountain ranges (1). Opposing forces exerted on the surrounding landmass sees a clean, brittle fracture dividing the landmass for the most part (2, 3), but along the fold there may be instances of little to no preexisting fractures. In these instances, the ductile crust begins to stretch forming an isthmus.*

There are two possible eventualities should the opposing landmasses separate significantly. If the isthmus stretches too thin it will break and form opposing tabs (4a) as seen on Kamchatka.

If secondary fractures give way on the opposing landmass, the isthmus will remain intact, but extract and drag a portion of the opposing range along with it (4b) leaving behind a notch in the landscape. The resulting tab displays the characteristics of the tabs seen in figure 91; ductile side arcs (A) with signs of stretching and thinning between (B), and a wide mountain or hill fragment across the end (C).

their differences is the key to locating mating landmasses. If an isthmian stub is characterized by a low elevation and a narrowing out to its furthest extremity, its mate would be similar in appearance, as the stub is clearly the product of an isthmian ductile fracture. But if the isthmian stub exhibits a band of ductile stretching in the form of side arcing (Fig. 91 A), with a low-lying strip of land between (Fig. 91 B) and a raised formation lining the end of the stub (Fig. 91 C), it represents isthmian extraction. We should instead search for its mate by identifying a notch of similar width in the opposing coastline.

Meanwhile, we have one more peninsular formation to review. While the Alaska Peninsula refers to a thin strip of land stretching out from Alaska's western coast along the Aleutian Arc, it should be noted that most of Alaska is itself a peninsular formation. Geologists attribute Alaska's geological origins to a range of theories, but they all seem to maintain that the peninsular mass is composed of multiple converging plates.

Figure 94. *Alaska folded back along the coast of Canada. The Canadian coast still retains the same Z-shaped coastline as Alaska, although it has fragmented substantially since the two separated.*

It is much more probable that, like the Asian peninsular formations, Alaska always existed as a singular structure that has fractured and is pulling free from the North American mainland. Similar to those instances, if we were to rotate the peninsula clockwise, we find it aligns with a conforming coastline in the Northwest and Nunavut Territories of Canada (Fig. 94).

One of the most intriguing aspects when we consider the peninsulas discussed thus far is that they all seem to be part of a linear pattern of fractures lining the northwestern Pacific Ocean. Figure 95 displays these peninsulas and their associated island arcs as they appear on a globe. When you view the arcs in this manner, you can easily recognize the consistent orientation of the peninsulas and their original continental fit, all having pivoted outwards in the same counterclockwise direction. The consistent pattern of unidirectional fragmentation belies the current theory, which maintains that the formations were derived from multiple inconsistent geological events. That theory proposes that some peninsulas erupted up from the seafloor, Kamchatka and Alaska among them; some are the product of randomly drifting and colliding

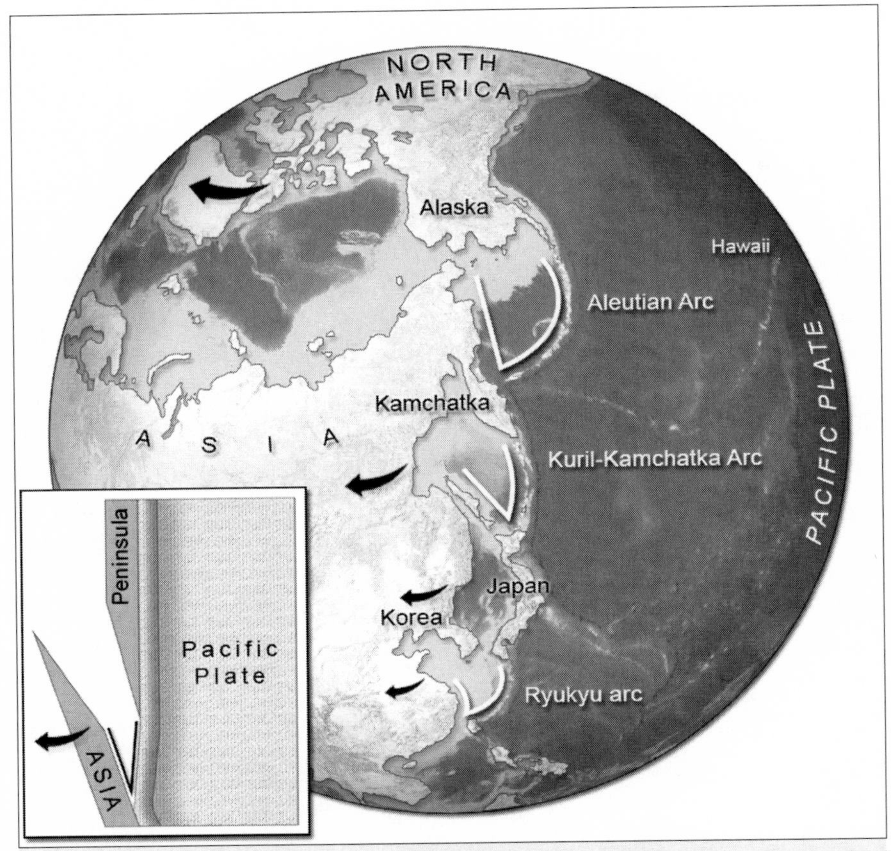

Figure 95. *Not only do the adjacent conforming coastlines that accompany each of these peninsular formations appear to support a consistent, unified theory for peninsula creation, but also associated with these formations are the downward extending expansion wedges we would expect to see for fragments bonded to the Pacific Plate and cleaved from the mainland. The inset image of canvas and a splintered frame once more illustrates the simplicity of this dynamic. This now seems far more reasonable than the current plate tectonic view which maintains that this consistent pattern came together in chaotic fashion with multiple wayward fragments colliding together to form Korea, volcanic upwelling creating Kamchatka and possibly Alaska, and Japan simply fracturing and separating from the mainland.*

continental fragments, as in the case of Korea; and some fragmented off from the mainland, such as Japan. But are these really the most logical explanations?

When we examine geological structures offered as proof of

colliding continental fragments, the evidence appears to support a more consistent pattern of crustal fracturing and separation. Along with the previous evidence that seafloor trenches in the Pacific formed by seafloor folding and not subduction, the evidence argues strongly in favor of the Earth expansion theory.

Deconstructing the Atlantic

Armed with the ability to recognize continental fractures and their direct effect on seafloor crust, we can apply this knowledge to understanding the true origins of other geographic structures throughout the globe. Let us begin in the Atlantic.

We know that South America was at one time directly attached to the African continent. The current configuration for this fit, displayed in Figure 96, depicts South America pocketed into an accommodating African coastline, while its tip extends southward roughly a thousand miles beyond the tip of Africa. When we take into account the coastal stretching and ductile fracturing occurring all along the Argentine coast, the disparity in length becomes far less significant.

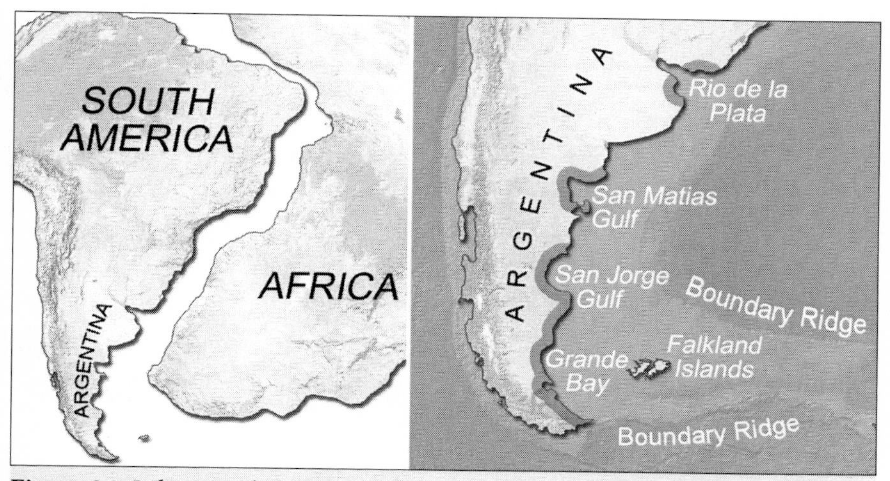

Figure 96. *When South America is pocketed into the side of Africa, its original fit, the South American continent extends 1,000 miles beyond the tip of Africa. Multiple ductile bays along the Argentine coast account for this disparity.*

Grande Bay, in Argentina's southernmost region, is a very clear ductile fracture exhibiting the signature arc and accompanying expansion wedge, complete with boundary ridges extending from its upper and lower cusps. It is also important to point out that the Falkland Islands, currently believed to be a fragment of land broken free of Africa, are centered in the expansion zone of Grande Bay. This suggests they may have been created by magmatic upwelling along the fracture's central divergent boundary, much like Iceland formed on the Mid-Atlantic Ridge. The Nastapoka arc, located on the southeastern shore of Hudson Bay, Canada, which many had once erroneously believed to be a section of an impact crater, is a ductile fracture arc that exhibits this same geographic feature set, with the Belcher Islands in the center of its arc (Fig. 97).

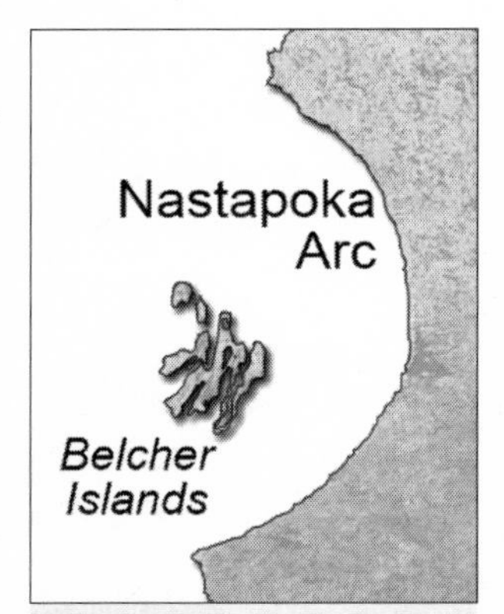

Figure 97. *Nastapoka Arc in the Hudson Bay is another example, like the Falklands, of a ductile fracture with an island formed by magmatic upwelling on an expansion zone.*

Suturing the Grande Bay ductile fracture back together would reduce the 1,000-mile overlap between South America and Africa by 445 miles. Further north, the San Jorge Gulf, San Matias Gulf, and Rio de la Plata would have likewise been closed while attached to the African continent, reducing the overlap by another 346 miles. This changes the discrepancy to a negligible 89 miles, which can be accounted for by ductile stretching of the surrounding coastline during fracture.

This stretch of torn coastline, along with a distance between the two continents that increases as we move southward, suggests that South America was methodically ripped or peeled away from the African continent from the south. It ripped free beginning at the southernmost point, stretching in stages, while the remainder of the continent above the Rio de la Plata proved the last section of resistance. Finally, the continents underwent brittle fracturing to break cleanly into two and drifted apart into the configuration we see today.

Many Rivers Flow Along Continental Fractures

Significant to our earlier Atlantis hypothesis is the fact that the Rio de la Plata is the product of a ductile fracture. Further stress on the region has generated brittle fracturing, which extends off the river up toward the Paraná Delta. According to Plato's *Critias,*

> *[Poseidon], being a god, found no difficulty in making special arrangements for the centre island, bringing up two springs of water from beneath the earth, one of warm water and the other of cold ... fountains, one of cold and another of hot water, in gracious plenty flowing.*[70]

Some have suggested that the city of Atlantis might have sat upon a geothermal site that provided an endless supply of hot water. The Rio de la Plata comprises a combination of fractures of varying forms that render the region highly susceptible to geothermal activity. The first discernible fracture is at the mouth of the bay, where a ductile fracture exhibits itself in the form of two halves of a well-defined arc. This represents the original arced bay that appeared as the Rio de la Plata was forming. Extended stress on this region would later generate a brittle fracture, creating the remaining western region of the Rio de la Plata, effectively splitting the original ductile arc in half.

We need only revisit the Kamchatka region to see multiple examples of combination-ductile-to-brittle fractures (Fig. 98). Typical examples of secondary fractures off the back of a ductile fracture can be observed in three locations. The first is a brittle fracture extending through the center of a ductile arc, as already seen in the Rio de la Plata (Fig. 98 top).

The next instance is a symmetrical fracture where two mirrored fractures open within a ductile fracture to each side of the center. Figure 98, center, depicts an instance of this at the Gulf of Anadyr. It is here, along with the Rio de la Plata, that we realize an important trend. Major riverways like the Paraná, Uruguay, and Anadyr are extended brittle fractures—creating low points through which mountain runoff

---

[70] *Critias* 113e,117a

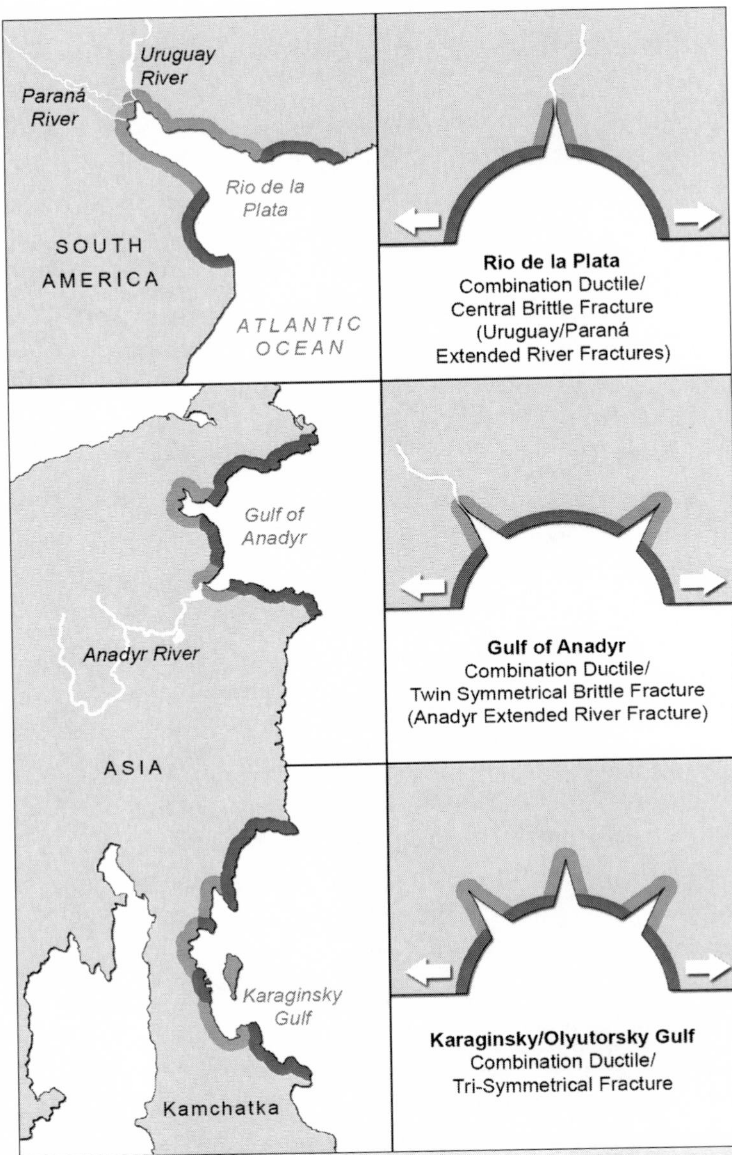

Figure 98. *Three common types of combined fractures. Simple central brittle fracture extending from a ductile fracture (top) forming the Rio de la Plata. Twin symmetrical fracture (center) forming the Gulf of Anadyr. Tri-symmetrical fracture (bottom) forming the Karaginsky/ Olyutorsky Gulf.*

has naturally flown down and through—as opposed to random paths defined by long-term erosion.

The last instance of a ductile-to-brittle fracture is the combination of central and side fractures that creates symmetry across the span of the initial ductile fracture (Fig. 98 Bottom). The Karaginsky/Olyutorsky Gulf provides an excellent example of this type of fracture. As we will soon observe, there is at least one instance elsewhere on the planet where the side fractures are two to a side versus one, as seen here.

As we move further north along the South American coast, we encounter a ridge formation extending out from the Brazilian coast that is mirrored by a ridge extending from southwestern Africa (Fig. 99). Within the plate tectonics model, these formations are understood to be the product of hotspot activity. The theoretical hotspot, Tristan, happened to sit on what would become the Mid-Atlantic Ridge, the divergent boundary lying between the continents. As the continents separated, Earth's shell (composed of the continents and new seafloor crust) rotated northward; the hotspot, along with the rest of Earth's core, remained in place. This created mirrored downward angled ridge

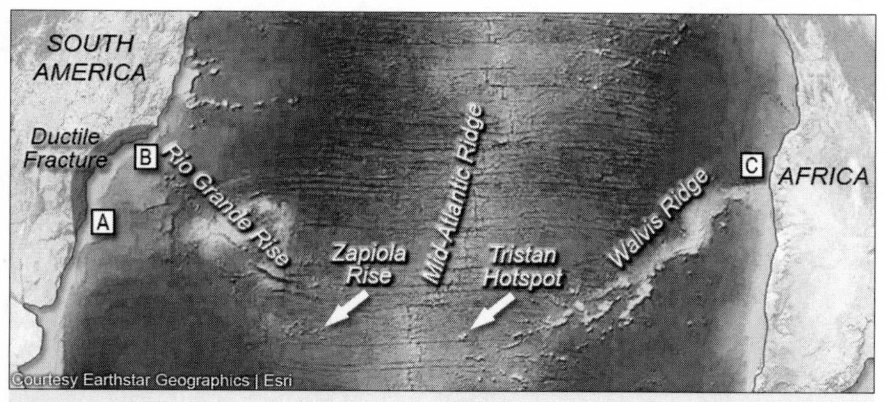

Figure 99. *The Rio Grande Rise and Walvis Ridge, extending from the South American and African coasts respectively, are currently regarded as hotspot formations linked to the proposed Tristan hotspot. This requires that the continents separated precisely over a hotspot with the Mid-Atlantic Ridge riding directly over it for millions of years to create the mirroring effect. Yet once again, we see ductile fracturing linked to the features. Along the Brazilian coast, a wedge-shaped rise centers on and extends out from the extremities of a ductile arc (A, B).*

paths traversing the Atlantic. These angled ridge formations, according to this theory, consist of the Rio Grande Rise extending out from Brazil and the Walvis Ridge extending out from the coast of Africa.

This view, however, overlooks the fact that the Rio Grande Rise is the section of a ridge that once again aligns with the cusp of an arced coastline (Fig. 99b). The slight rise extending from the arc's southern cusp (Fig. 99a) defines the lower boundary of a raised expansion wedge that spans the width of the arc. This is the uppermost arc in the series of ductile fractures extending up from the continent's southern tip (Fig. 96), and it was the last major ductile fracture to form on the continent before the remainder of South America and Africa fractured completely through to the north. The Walvis Ridge and Rio Grande Rise form the downward expansion wedge resulting from this final fracture.

The initial expansion wedge began to form as the southern half of

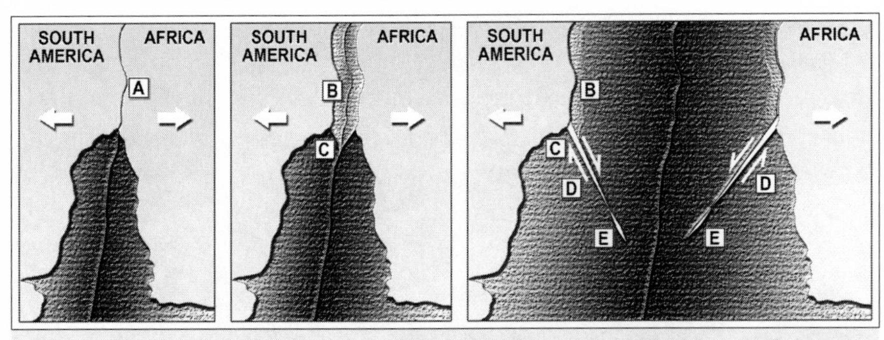

Figure 100. *The Rio Grande Rise and Walvis Ridge originated when South America and Africa separated in the north (A). The suddenness of the brittle fracture, which saw the creation of new seafloor (center B) catastrophically expanding more rapidly than the preexisting seafloor to the south, likely caused a V-shaped split in the Mid-Atlantic expansion ridge (center C). In turn, this division between seafloor crust in the north and the south continued with the ongoing extension of twin fractures (D), maintaining instability between plates to the north and south. As the continents continued to separate, pulling the Atlantic seafloor along with them, the fractures extended further and further in the prevailing southerly direction (E), traversing the lateral fracture zones and flow lines emanating from the Mid-Atlantic Ridge. In order for this to be the case, magma rising up through these fractures could not be allowed to harden and fully bond, stabilizing and melding the plates together. This would suggest a period of catastrophic plate movement of feet or miles per day—versus the current rate of inches per year—which would prevent the fractures from bonding.*

South America peeled away from Africa, this being where the Brazilian ductile fracture arc is located. Shortly thereafter, the two continents broke free of each other in the north, and a new expansion zone (or wedge) began to appear in the form of the Rio Grande Rise and Walvis Ridge.

According to this new view of continental fracturing, the upper and lower cusps of the Brazilian ductile fracture arc, which created the Rio Grande Rise, were once shared points; these in turn shared a third common point where the Walvis Ridge intersects the African coast (reference Fig. 100). As noted, this geographic point demarcates a period of ductile separation of the continents to the south and a sudden final brittle fracture to the north. Supporting this view is the pronounced difference in continental shelf formations beneath these ridges, where they are unbroken and consistent in width, in contrast to the markedly reduced shelf to the north.

Just to the north lies the next geographic feature we will review, the Bight of Biafra. The Bight of Biafra has all of the earmarks of a ductile fracture. This arc of land cut into the side of the African coast also exhibits pronounced cusping at its southern point. The arc is bisected by a feature known as the Cameroon line, a chain of linear volcanoes

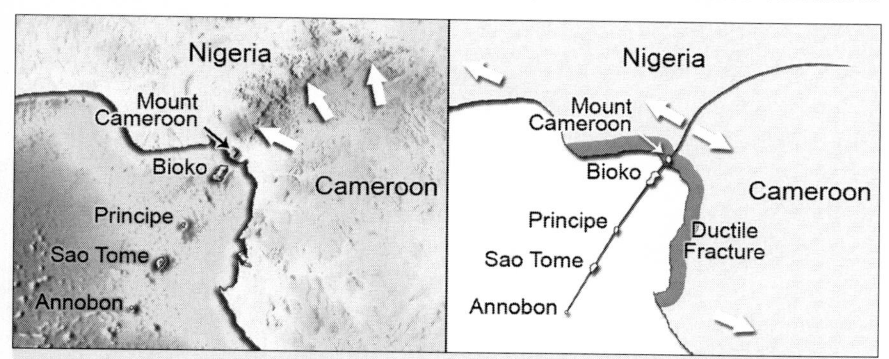

Figure 101. *The Cameroon line is a chain of volcanoes extending from the Atlantic seafloor and onto the African continent, bisecting the Bight of Biafra. Many believe it is the product of the Cameroon hotspot cutting a swath through the two crusts. More likely, the Cameroon line is centered on a ductile fracture arc and represents an inland extension of the fracture in brittle form. Instability on either side of this inland fracture affects the attached adjacent seafloor, causing a similar hairline brittle fracture to form and extend out onto the Atlantic seafloor.*

that includes the islands of Annobon, Sao Tome, Principe, and Bioko rising up from the Atlantic seafloor and Mount Cameroon lying seven miles inland (Fig. 101 left). The line extends inland beyond Mount Cameroon and is marked by further volcanic activity forming a highly discernible eastward arcing ridge.

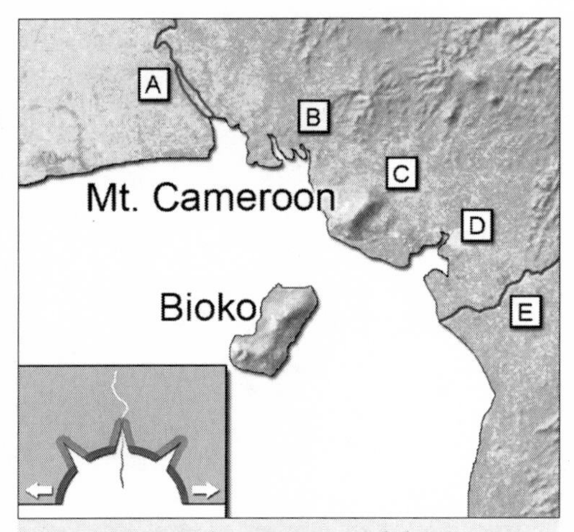

Figure 102. *A close-up view of the Cameroon line where it intersects the coast. The symmetry of the fractures on each side of the Cameroon line clearly signifies that the Cameroon line is the central brittle fracture within a multi-symmetrical fracture (inset from Fig. 98). Note the link between brittle fractures and inland waterways as demonstrated by the outer two fractures A and E.*

While plate tectonics credits hotspot activity for much of the formation, the evidence overwhelmingly points once again to fracturing of the African continent. Similar to the Rio de la Plata (Fig. 98), the fracture comprises an arced ductile fracture and a central brittle fracture extending further inland (Fig. 101 right). Unlike in the case of the Rio de la Plata, the brittle fracture has occurred without significant separation, although sufficient activity has extended it out onto the adjoining seafloor. Magma has welled up through this fracture to form the multiple volcanic islands and inland volcanic range.

In further support of the fracture theory, the Cameroon line not only bisects the arced Bight of Biafra but to either side of this intersection lie secondary fractures (Fig. 102 A, B, D, and E). As we have seen earlier, this is evidence of extended stress and stretching within a ductile fracture. The symmetry of the two fractures nearest the Cameroon line (B and D) is quite remarkable in relation to the central brittle fracture (C), while the outer two fractures (A and E), though differing from each other in appearance, demonstrate the link between fractures and the natural flow of rivers through them.

It seems highly unlikely that a hotspot lying deep beneath the African continent randomly cut a path not only down the middle of an arcing coast, but also directly down the middle of these outer fractures, to generate this pattern of symmetry. Associating these surface features with ductile and brittle fractures, on the other hand, is consistent with very basic rules of material properties and fracture mechanics. It also eliminates the need to base our theory on an invisible and unproven force lying beneath the planet's surface.

Moving north up the Atlantic, thorough analysis of continental formations will prove crucial in solving a large and complex structure linking the Americas: Middle America, which comprises Mexico, Central America, Venezuela, and the Caribbean Islands. One of the earliest hypotheses was put forth by the same man who introduced the hotspot theory, J. Tuzo Wilson. Keeping with the plate tectonics view of randomly drifting continents and continental fragments, Wilson proposed that the stretch of land linking Mexico to the South American continent was mostly the product of continental fragments in the Pacific having traveled east and wedged themselves between the two continents.

This view maintains that the Caribbean Plate lying in the Pacific existed as a small plate surrounded by a few continental fragments

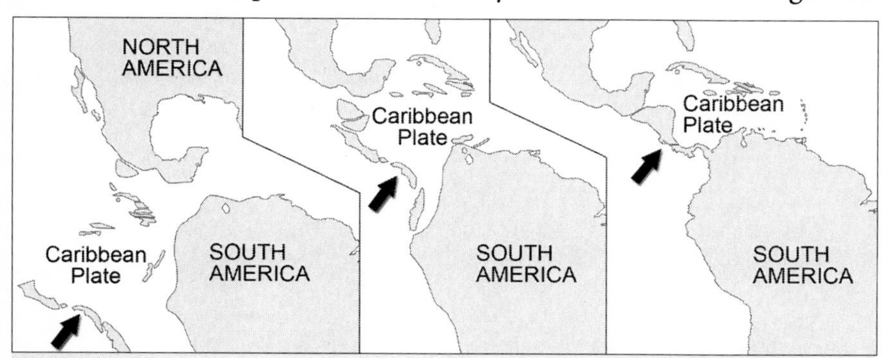

Figure 103. *J. Tuzo Wilson, the same geophysicist who proposed the existence of hotspots, is also responsible for the theory that the Caribbean Plate navigated its way up from the Pacific, between the separating Americas and on to its current location. The theory clings to the basic element of plate tectonics—the idea that continents and continental fragments are randomly drifting in all directions.*

(Fig. 103 left). These fragments were collected from various continents during hundreds of millions of years of random drifting. The plate eventually drifted up and through the gap separating the Americas (Fig. 103 center) with the plate's western extremity failing to clear, wedging itself between the two continents. Subterranean forces propelling the plate eastward caused fragments lining this western backside edge of the plate to be compressed together within the relatively small gap, thus forming a continuous chain of land: Central America (Fig. 103 right). Much like the proposed origins of Korea, it is only through the fortunate timing of our existence with perfectly coordinated crustal plate movements that we witness these fragments today fully merged into a land bridge.

Patterns along the Atlantic seafloor, however, reveal alternate origins. Throughout the Atlantic, fracture zones lie in parallel, with a few exceptions—one being in latitudes shared with the Caribbean Plate. Two fracture zones, the Fifteen-Twenty and Sierra Leone, converge near the coast of Africa (Fig. 104 A). This convergence appears to mark the point where the Americas began their process of separation. The two fracture zones separate immediately and run parallel shortly thereafter

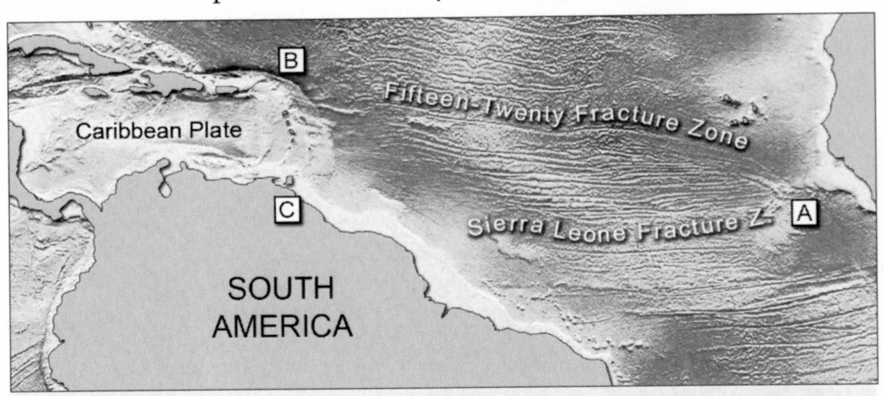

Figure 104. *The Fifteen-Twenty-Sierra Leone wedge. Whereas all other fracture zones in the Atlantic tend to run in parallel, the Fifteen-Twenty and Sierra Leone fracture zones converge just off the coast of Africa (A). At the opposite western end, the two fracture zones open up to the eastern extremity of the Caribbean Plate (B and C). This reveals the early split of the Americas shortly after separation from the African continent and ties it to the creation of the Caribbean Plate.*

forming the Fifteen-Twenty-Sierra Leone wedge. It reveals that the Americas separated north and south by 500 miles almost immediately after breaking free of Africa.

The precise point of separation of the Americas can be traced over to the Caribbean span, where the western ends of the Fifteen-Twenty-Sierra Leone wedge extend toward the upper and lower extremities of the Caribbean Plate (Fig. 104 B and C respectively), which retain the wedge's central width of just over 500 miles. While the lower portion of the wedge links to South America where the Caribbean Island Arc intersects the coast of Venezuela, the upper portion of the wedge aligns with the upper extremity of the Caribbean arc. If we continue to follow the upper portion of the island arc westward via the Dominican Republic and Cuba, we come to the Yucatan Peninsula—and a feature which offers clues to the region's true origins.

Extending out from the Yucatan peninsula to the west and north and stretching from the Caribbean arc in the east to the coast of Mexico in the west is a stepped continental shelf (Fig. 105 A). At the op-

Figure 105. *At each end of the Caribbean Island Arc lie two stepped continental shelves that share similar dimensions. The Yucatan shelf (A) had originally been pocketed into the Guyana shelf (B), and the Yucatan coast (C) joined to the Honduran coast (D); they, along with the Isthmus of Panama, were peeled from the top of the South American continent. The Caribbean wedges that define the eastern edge of the Caribbean Plate reveal that much of the Venezuelan coast is ductile. Compressed back to its original size, the reduced coastline allows Honduras and the Yucatan to reach and fall along the Guyana coast.*

posite end of the Caribbean Island Arc, extending off the coast of Guyana, lies a similarly stepped continental shelf (Fig. 105 B). The Guyana shelf looks as though a large portion of it has been plucked free; the Yucatan shelf appears to be that missing piece. In conjunction with the alignment of the Fifteen-Twenty-Sierra Leone wedge, it reveals the possibility that the Yucatan and the rest of Central America were peeled from the top of South America, opening up a void that would become the Caribbean Plate.

Figure 106. *Yucatan and Honduras exhibit opposing coastlines of similar length and conforming shape.*

The Caribbean arc is the recorded path of the Yucatan as it broke free from the upper portion of South America and rotated up and away some 2,000 miles to its current location. The long trip likely accounts for some slight distortion in the fit of the two shelves. Keeping with the theory of Earth expansion, it seems that the Caribbean Plate and Central America did not form through random drift and collision but through separation.

Honduras, which is widely believed to be a random continental fragment that collided with the Yucatan long ago, is instead the product of fracturing from Yucatan as the entire region separated from the South American continent. The opposing coasts not only measure roughly the same 400 miles in length, but the coastlines also have conforming contours; the Yucatan with its slightly concave extremity and Honduras with a similarly matching convex extremity (Fig. 106).

Meanwhile, three arced ridges draping down off the Caribbean arc and spanning the coast of Venezuela form expansion wedges, revealing that significant ductile expansion occurred across much of South America's northern coast (Fig. 105). Compressing this northern coastline back to its original width would allow the thin layer of mass known as the Isthmus of Panama to lay across the top of the South American continent. The combined mass of the Yucatan and Honduras would fall into the northeastern notch of the continent in the vicinity of Guyana. This reconstructs the original consolidated continent of Central and South America. The current state of separation leaves South America barely tethered to the North American continent.

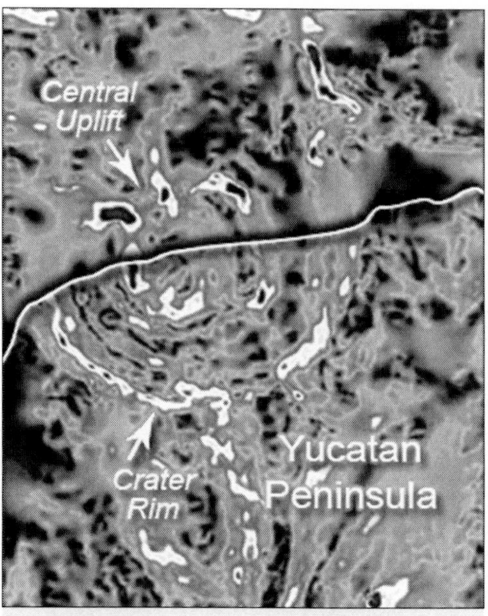

Figure 107. *Gravity map of the Yucatan Peninsula revealing the presence of a crater lying beneath the peninsula's surface: the Chicxulub impact crater.*

There may be a method of confirming this theory of origin: an impact crater lying at the northwest coast of the Yucatan Peninsula, believed by many to be tied to the extinction of the dinosaurs.

Chicxulub lacks all visible physical features of an impact crater. Researchers discovered this subsurface impact crater using gravity maps (Fig. 107). Gravity maps reveal fluctuations in Earth's gravity field, which are directly associated with mass that can vary significantly from one geographic feature to another. On the Yucatan these fluctuations reveal a 112-mile wide circular pattern, evidence of an impact crater that has been buried deep beneath Earth's surface.

Of interest here is the rim's limited extension beyond the Yucatan coast. Much of the outer concentric arc abruptly truncates at the coastline. This suggests that the other half of the outer rim may lie elsewhere. When Chicxulub slammed into Earth, naturally occurring radial fractures would have formed, extending out from the impact's center.

One of these fractures likely explains the Yucatan's linear coastline, set closely along the arc's center, while also potentially explaining the catalyst for the separation of the Americas into two continents. This assumes that the impact occurred completely inland in the vicinity of Guyana on a unified North and South American continent, where the Yucatan remained merged with the coast. The fracture generated by the impact left Guyana and the Yucatan merged but fragile. When the Americas migrated away from Africa and Europe, the fracture eventually succumbed to the stress of the movement, resulting in the initial separation of the Americas and sent the Yucatan off on its long migration away from Guyana. If this were true, gravity maps of the Guyana coast may reveal the existence of Chicxulub's other half and corroborate the true genesis of Central America.

Chicxulub's second half likely resides along a 110-mile coastal stretch extending between the mouths of the Courantyne and Essequibo Rivers (Fig. 108). The similarity in width between these rivers and Chicxulub on the Yucatan may be mere coincidence. But the location certainly provides a conforming corner for the Yucatan's northwest corner to nestle into. Meanwhile, the two rivers' origins could be linked to the crater's rim, similar to the way in which several *ceynotes*, or sinkholes, have formed along its rim on the Yucatan.

Nestling Central America and the Yucatan back atop South America

Figure 108. *Depiction of the Yucatan placed back along the South American continent, with semicircles depicting the location of Chicxulub on the Yucatan and my proposed location of Chicxulub's other half.*

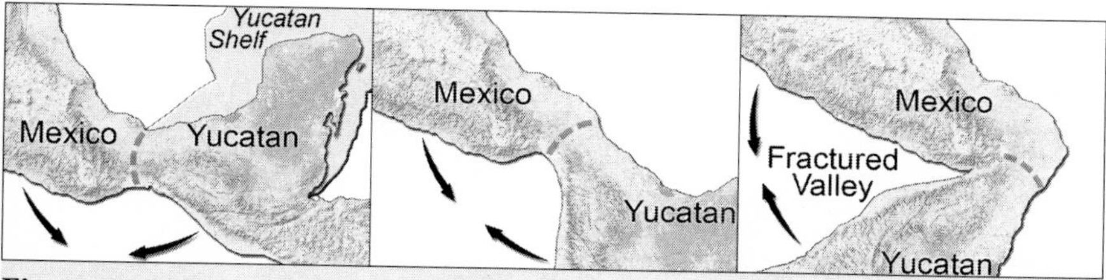

Figure 109. *Lying between greater Mexico and the Yucatan is a transition point between two mountain ranges, a narrow neck of land known as the Isthmus of Tehuantepec. Here the land above and below was originally folded back upon itself.*

provides a base for the next layer of separation. We find this next layer in southern Mexico, just north of Central America, where two mountain ranges—the Sierra Madre Del Sur Mountains in the north and the Sierra Madre de Chiapas in the south—meet. The region not only forms a narrow lowland neck, the Isthmus of Tehuantepec, but it also happens to align with the lower step of the Yucatan shelf, which marks the breakaway point from Guyana (Fig. 109 left). Since the stepped shelf was originally connected with the farthest point south on the Caribbean wedge, the only options for placement of the land above are to the north and possibly the west. Therefore, the isthmus marks the vertex between the western coasts of Mexico and Central America, which were originally folded one upon the other (Fig. 109 right).

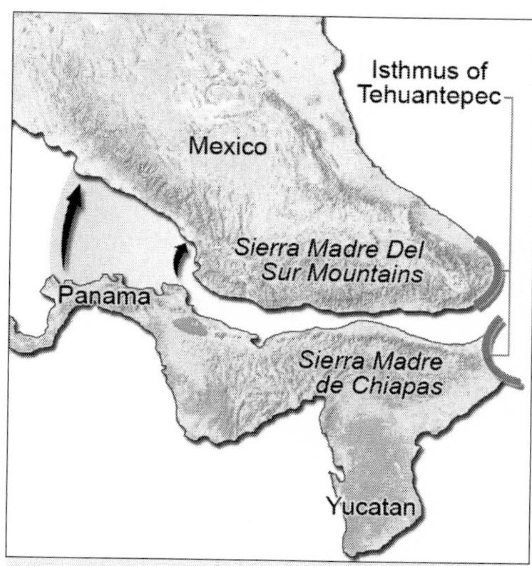

Figure 110. *The coastal conformation extending either side of Isthmus of Tehuantepec.*

We have previously demonstrated the weakened state of continental crust along the downward fold or valley between mountain ranges; this is another instance of a brittle fracture occurring within a valley. When the mountain-lined coasts to the north and south of the Isthmus of Tehuantepec are brought back together, we can reconstruct the ancient valley between them.

Rotating the stretch of land south of the isthmus clockwise not only brings the two mountain-lined coasts together, but finds the protrusive coastline extending from Costa Rica to Panama nesting in a conforming pocket of Mexico's western coast just north of the Tehuantepec Isthmus (Fig. 110).

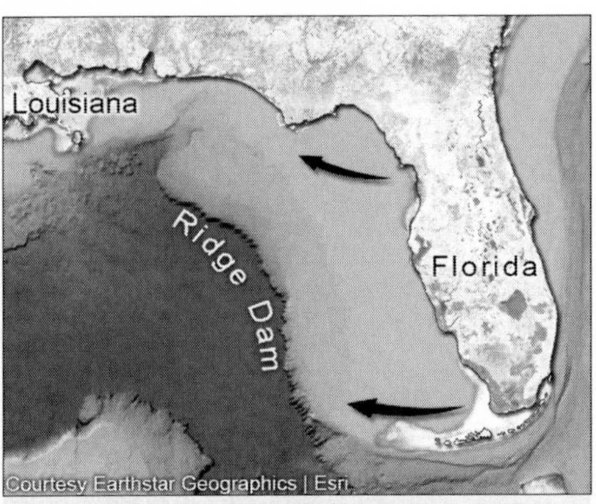

Figure 111. *Like peninsulas analyzed earlier, once again we see a chain of islands and a ridge extending from Florida's tip back to the mainland making it clear that this peninsula likewise was once nestled into the mainland.*

The next level of reconstruction comprises two parts. The first revisits our findings on peninsular formations and applies them to the Florida peninsula. Once again, we see a linear path of islands extending from the end of the peninsula (Fig. 111). These islands are the exposed portion of a ridge that stretches back to the mainland toward the Mississippi River; a conforming pocket lies along the coast, extending the length of the Florida panhandle capping off at the end by the protruding coastline of Louisiana. This ridge extending from the tip of Florida borders the seafloor that formed as the peninsula pulled free of the continent—a seafloor that eventually became another sediment-filled continental shelf.

The next part of the reconstruction also originates at Louisiana and the Mississippi, this time extending west across the southern coast of Texas and down to the Isthmus of Tehuantepec. By now, this large coastal feature should be readily recognizable. The whole stretch of coastline is one of the globe's largest natural arcs and is most certainly a ductile fracture.

The span of the arc stretches roughly 1,200 miles and, like the Bight of Biafra we find similar signs of a brittle fracture traversing the coastline near the ductile fracture's center. The brittle fracture exists

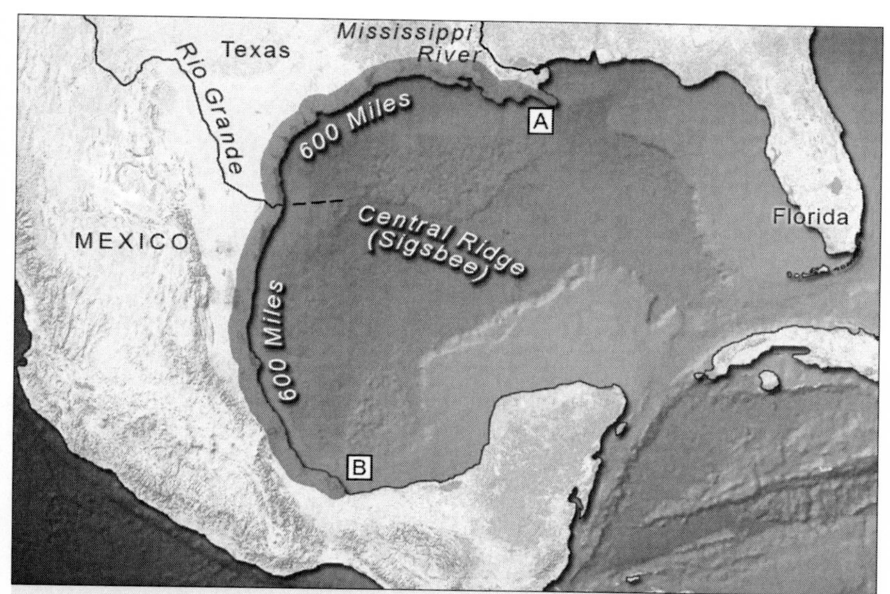

Figure 112. *The Tex-Mex ductile fracture arc (highlighted) extends 1,200 miles from the tip of Louisiana in the north (A) to the ancient bend at the Isthmus of Tehuantepec in the south (B). Similar to the Cameroon line (Fig. 101), a central brittle fracture extends inland and also out onto the seafloor, in this instance, in the form of the Rio Grande inland and as the lateral edge of the Sigsbee Escarpment in the Gulf of Mexico.*

inland in the form of the Rio Grande, the major river forming the border between Texas and Mexico (Fig. 112). It extends out onto the adjacent seafloor, forming an aligned fracture-generated ridge called the Sigsbee Escarpment. This linear formation lies 600 miles from the tip of Louisiana and 600 miles from the Isthmus of Tehuantepec, the heart of this expansive Tex-Mex ductile fracture.

Scientists currently recognize that a rift exists along a portion of the Rio Grande; a large section of the river defined by several deep basins extending from Chihuahua, Mexico to central Colorado has been designated the Rio Grande Rift. Now we can identify the river's path to the Gulf of Mexico as an extension of this rift. The gulf itself was formed when the Rio Grande Rift ripped open along a ductile coast, much like the Bight of Biafra ripped open along the Cameroon fracture.

Rolling back the geological clock and closing the Gulf of Mexico nests the Isthmus of Tehuantepec and tip of Florida up against the Mississippi River. The separation of the two at the Mississippi River is likely no coincidence, as this river too is likely the site of a brittle fracture. We do not see the extent of visible rifting found on the Rio Grande, but as in that case, scientists do recognize that a portion of the river lies on a rift called the Reelfoot Rift. Reelfoot in turn lies within the infamous New Madrid Seismic Zone. The New Madrid zone caused a devastating series of quakes between 1811 and 1812—including, over a three-month period, four of the country's most devastating recorded quakes.

While scientists believe the rifting is limited to a region of the Mississippi extending 150 miles between Missouri and Arkansas, the Floridian pullout suggests that, similar to the Rio Grande Rift, there is a rift extending all the way to the Gulf of Mexico. The Floridian pullout is one of the last in a sequence of events that occurred during the forming of Middle America.

Figure 113 approximates Middle America's formation through a series of images. The sequence of events begins with North and Middle America breaking free of South America between the Yucatan and Guyana. The stress applied to an ancient radial fracture within the Chicxulub Crater sees the Yucatan and Honduras pulling free of the southern continent. Honduras, not having been as severely fractured by the Chicxulub impact, initially resists separation but succumbs, fracturing from the Yucatan shortly afterward.

This pull on the South American continent also began expanding the northern coast of the continent as it became exposed, thinning out the coastal region and stretching it over 500 miles to create most of what we know today as Venezuela.

Next the Rio Grande Rift, now home of the Rio Grande River, began to open up at its ductile mouth, forming the Gulf of Mexico. Then the Florida peninsula, anchored to the Atlantic seafloor, pulled free of the North American continent. Meanwhile, the Yucatan's arcing path upward and westward opened a void between itself, the Atlantic, and the South American continent, creating the Caribbean Plate. As this happened, volcanism along the two plates' borders created the trail of Caribbean islands.

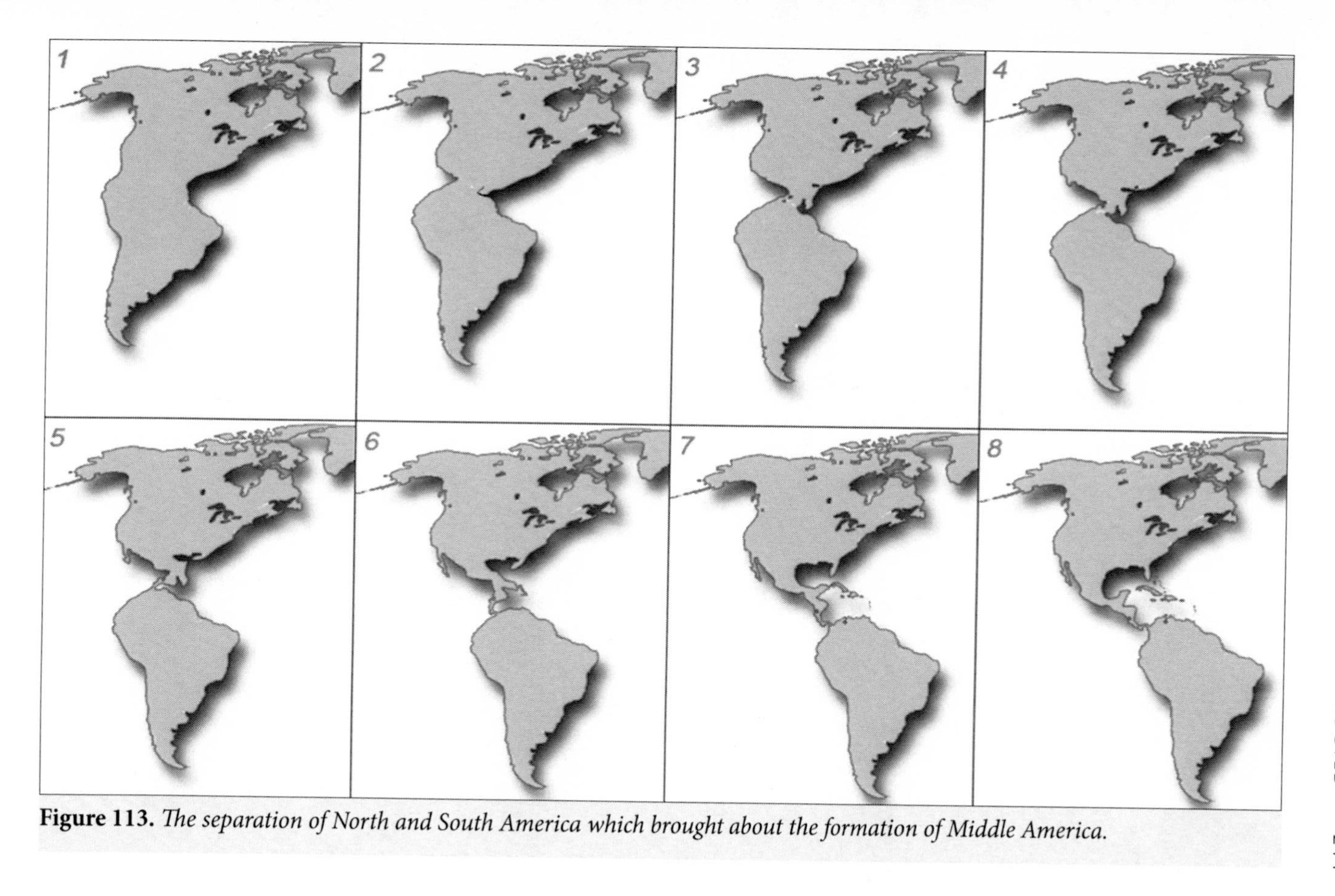

Figure 113. *The separation of North and South America which brought about the formation of Middle America.*

As the gulf opens further and North America continued to drift northward, the Panamanian Isthmus pulled free of Colombia. The continuing upward pull rotated it counterclockwise until it reached its current state, with the North American continent remaining barely tethered to South America by this narrow stretch of land.

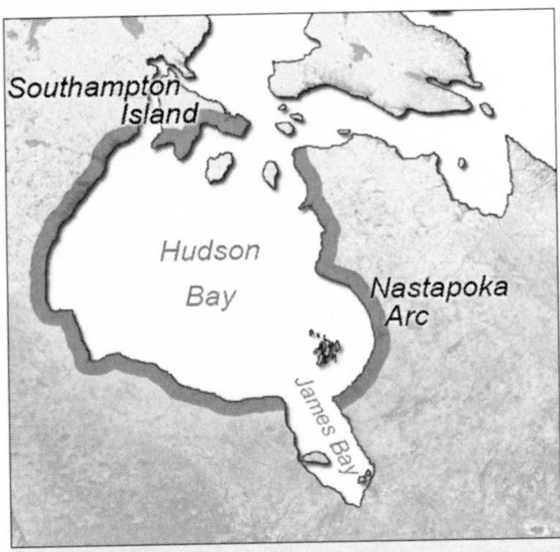

Figure 114. *Hudson Bay is one large ductile fracture. A brittle fracture has separated the western cusp, Southampton Island, from the mainland, but it still extends out toward its sister cusp in the east.*

Let us move northward to Canada to examine our next geographic structure: a large body of water known as Hudson Bay (Fig. 114). The whole of Hudson Bay is a compound ductile fracture, very similar to the Bohai Sea (Fig. 88). Like Bohai, secondary fractures extend off the main fracture in the form of James Bay in the south and a circular arc just above it along the eastern coast of the Hudson, known as the Nastapoka arc.

Scientists have concluded that this arc and the Belcher Islands, which together appear at first to be the product of a terrestrial impact, are merely an arc-shaped brittle fracture, with the Belcher Islands' positioning at the arc's center the product of happenstance. What cannot be overlooked, however, is the fact that similar formations exist in the Falkland Islands (Fig. 96), Karaginsky Island off the coast of Kamchatka (Fig. 98), and Bioko, which is centered on the Bight of Biafra (Fig. 102). This signifies a consistent relationship between islands centered off arcing coastlines. These arc-and-island sets bolster the theory that the islands are tied to continental ductile fracturing and the subsequent seafloor fracturing, which then contributes to island formation where magma has seeped up through central divergent boundaries.

As we move on toward Europe, note the transatlantic link between the Mediterranean Sea and the Saint Lawrence River. Fracture zones

Figure 115. *The Pico-East Azores Fracture Zone reveals the ancient connection between two continental fractures, the Saint Lawrence Gulf in North America and the Mediterranean Sea in the east, once a single continuous lateral fracture before the Atlantic opened up.*

extend between two points that were once merged at a central divergent boundary. Because of this we can determine that the East Azores and Pico fracture zones define the path of separation between the two bodies of water (Fig. 115). If we were to close the Atlantic and bring the continents back together, we would find Nova Scotia resting back along the Moroccan coast and Newfoundland coming to rest on the coast of Spain. Both of these regions, which currently lie over 3,000 miles apart, are similarly split, with the Saint Lawrence River splitting North America and the Mediterranean Sea separating Europe from Africa. The Saint Lawrence River is a brittle continental fracture and an extension of the Mediterranean Sea, which itself is a continental fracture. In contrast to the Mediterranean fracture, which exhibits an immense amount of expansion, the Saint Lawrence has experienced much less, with the exceptions of the Gulf of Saint Lawrence and the Great Lakes at its western extremity. It is clear that these bodies of water were originally a shared fracture before the Atlantic Ocean opened up and separated the continents. The true origins and significance of this shared fracture will be revealed as we delve deeper into Earth's geological past.

Meanwhile, within the Mediterranean, a very intriguing coastal feature lies along the northern coast of Africa. It is a rectangular rather

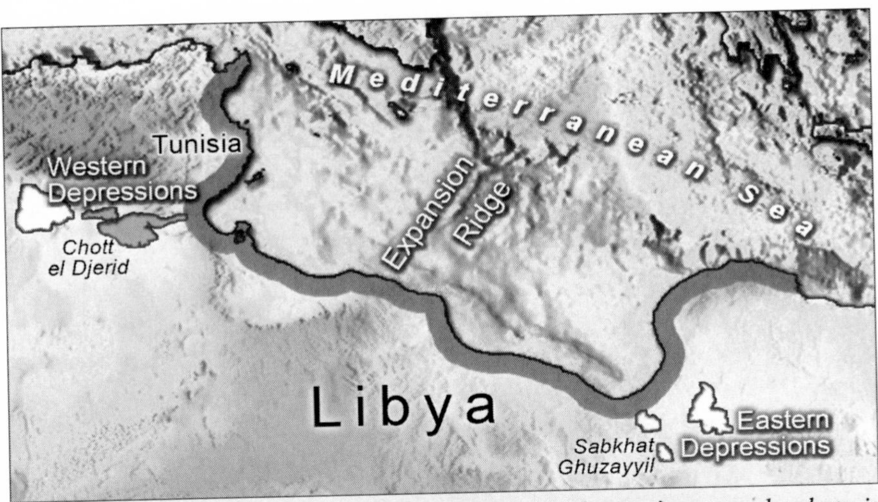

Figure 116. *The Libyan-Tunisian ductile fracture. The fracture's rectangular shape is unique to previous examples, but key identifying elements exist exactly where expected. The Medina escarpment forms a central expansion ridge, while there are signs of early-stage secondary ductile fracturing on the inside corners, where the thinning of the continental crust has created a series of depressions.*

than oval recess, but there is little doubt that it is a ductile fracture (Fig. 116). The opposing cusps either side of the recess provide the initial clue that it has been created by ductile tearing. A seafloor ridge known as the Medina escarpment, centered directly between the cusps and extending perpendicularly from the coast, provides the fracture's central expansion ridge, similar to what we saw with the Tex-Mex ductile fracture (Fig. 112). The similarity also extends to the rise in the fracture's center—although instead of forming a point, as in the Tex-Mex fracture, this rise is squared off. However, the final confirmation that this is truly a ductile fracture exists in signs of extended inland ductile stress: continental depressions positioned at each of the fracture's inside corners. The Tunisian lake Chott el Djerid lies at the western end of the fracture along with other smaller continental depressions, while on the eastern end there also exist multiple large depressions. These depressions are areas of the continent that dip below sea level. We have seen where extended stress on ductile fractures has led to secondary symmetrical brittle fracturing (Fig. 98, p. 196). In this instance, we have the onset of symmetrical ductile fracturing with thinning of continental crust occurring in the inside corners of the recess, resulting in surface depressions.

To further support the theory that this is a ductile fracture, there exists a ductile fracture lying along the coast of North Siberia that is nearly identical in appearance. The North Siberian fracture not only shares the shallow rectangular nature and large rounded cusping seen on the Mediterranean fracture; it also exhibits a similarly squared-off central coastal rise (Fig. 117). The fracture has also incurred a secondary brittle fracture in its western inside corner in the form of the Khatanga Gulf and River, which mimics secondary fracturing extending into the corners of other ductile fractures (Fig. 98, p. 196, Fig. 116). Meanwhile, similar to the center of the Tex-Mex fracture, the North Siberian fracture's central rise is the location of one of the world's longest waterways, the Lena River, a brittle inland extension of the coastal fracture.

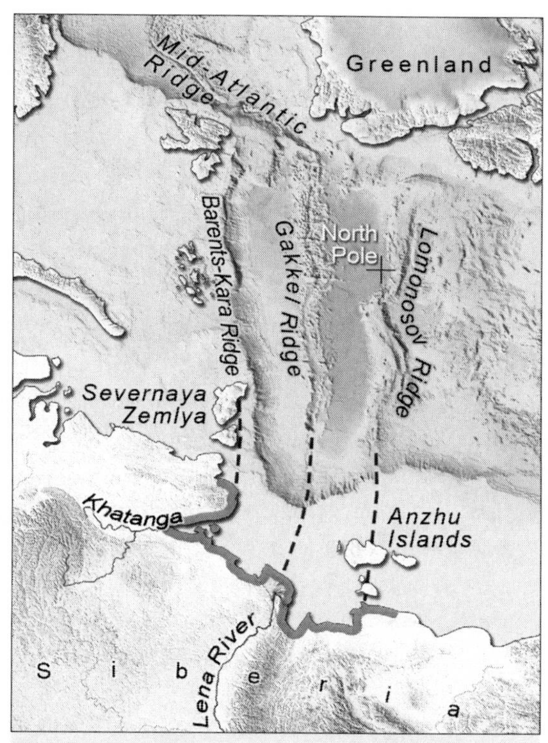

Figure 117. *The North Siberian ductile fracture. The perfect alignment of a trio of seafloor ridges provides some of the best evidence yet of the effects of a coastal ductile fracture on the adjacent seafloor. As in similar fractures, we see boundary ridges aligned with cusps, and we see the clearest evidence yet of the existence of an expansion zone, with the Gakkel Expansion Ridge extending directly from the fracture's central rise.*

Currently, the prevailing theory maintains that the mountains just east of the Lena River define a suture zone where plate boundaries are converging and thus compressing the plates together to form a mountain range. In fact, it is widely accepted that the region forming the eastern side of the suture is the western extremity of the North American Plate. But again in this case, plates are not converging, but fracturing and moving apart. In this instance, the northeastern portion of Siberia is and has always been attached to Siberia; what we are actually witnessing is the region slowly fracturing and breaking free of the Asian continent.

Once again, and in grand fashion, we see the very clear relationship between a continental ductile fracture and its adjoining seafloor (reference Figure 75, p. 163). First, we see outer boundary ridges extending from the opposing cusps. Aligned with the western cusp is the Barents-Kara Ridge, the leading edge of the Barents-Kara shelf, which forms the western boundary of the ductile fracture's expansion zone. The Lomonosov Ridge extends off the eastern cusp, forming the expansion zone's eastern boundary. And most significantly, centered between these two ridges is the Gakkel expansion ridge, the northern extension of the widely recognized Mid-Atlantic expansion ridge. The Gakkel Ridge terminates at a continental shelf, but if extended it clearly aligns down the center of the Siberian ductile fracture with the Lena River. This is perhaps the most visible evidence of a ductile fracture—having all three ridges aligned to the two outer cusps and the center of the fracture—but more importantly, the Siberian ductile fracture is linked with an established expansion zone.

A ductile formation of a slightly different sort lies to the southwest in the form of the Arabian Peninsula. In order to piece this portion of the world back together based on Earth expansion, Africa must rotate counterclockwise back along the Asian coast, but the existence of the Arabian Peninsula would initially appear to obstruct such a fit.

In the Indian Ocean, the Owen Fracture Zone, bisected by the Sheba expansion ridge, aligns to the bases of two mating 600-mile stretches of continental shelf, a recessed shelf along the coast of Iran-Pakistan and a protruding shelf along the coast of Somalia (Fig. 118 A^1B^1 to A^3B^2). This would clearly signify that these shelves and their corresponding coasts were once joined together prior to the Sheba Ridge expansion.

At the upper end of this fit are an array of flow lines lying at the bottom of the Gulf of Aden. Flow lines are lateral lines that are extruded

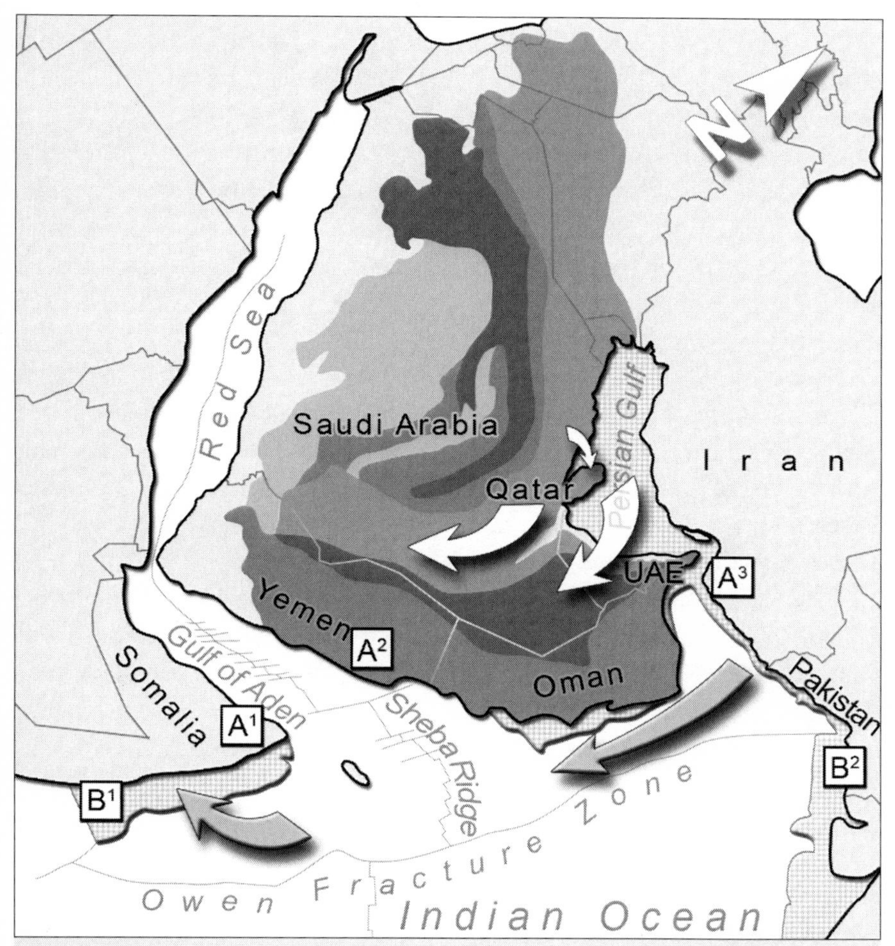

Figure 118. *Somalia (A¹-B¹) was once affixed to the coasts of Iran and Pakistan (A³-B²) before the African continent rotated free from the Asian continent. The Arabian Peninsula's origins appear to be revealed in its surface patterns, which maintain two linear alignments: one within Saudi Arabia running parallel to the Asian coast and the other within the UAE, Oman, and Yemen, reflecting their extraction from the Persian Gulf with the rotation of the African continent to its current position perpendicular to the Asian coast. A bend in the pattern records the transition between Africa's two alignments.*

out from a divergent boundary during expansion. Like fracture zones, they are mirrored striation patterns that mark the path of travel for two points separating away from a central point. The flow lines along the bottom of the Gulf of Aden reveal that Somalia was once attached to the coast of Yemen, joining point A^1 to A^2.

Since A^1 and A^3 are shared points based on mating continental shelves and A^1 to A^2 are shared points based on flow line alignment, logic dictates that all three points were once shared points along the Iranian coast. In order for this to have been the case, we have to account for Oman and the United Arab Emirates (UAE), which lie as obstructions directly between Somalia and Iran.

Surface patterns observed in satellite images of the Arabian Peninsula appear to have recorded the peninsula's transformation. The striated surface has become warped by the extreme ductile stretching and bending that formed the region. Figure 118 recreates those surface patterns that appear to confirm that Oman and the UAE were plucked out of the void that would become the Persian Gulf.

We can see that the Qatar Peninsula, having been pulled from a pocket along the Arabian coast, was affected by the same clockwise pull that occurred on a larger scale with the UAE and Oman. Between Qatar and the UAE exists a narrow light-shaded dagger-like pattern, extending inland and marking the separation between the peninsula's two distinct patterns of alignment. The linear pattern to the west remains aligned with the Asian coast, retaining the African continent's alignment as it originally was when it rested along the Asian continent. Meanwhile, the linear pattern immediately to the right lies rotated nearly 90 degrees from the Asian continent, suggesting that the region was extracted from the Persian Gulf along with the African continent as it likewise rotated clockwise to the same 90-degree angle away from the Asian coast. The pattern in its entirety appears to record the transitional bending and stretching action having occurred on the peninsula, marking Africa's transition from lateral to vertical.

Another piece of evidence that potentially supports ductile deformation of the region is linked to oil reserves. Saudi Arabia possesses some of the world's largest and perhaps best-known reserves—and according to this new analysis, it coincides with one of the planet's largest ductile deformations. The Gulf of Mexico is the site of the planet's largest ductile arc and is a major source of US oil production. Venezuela, a country with reserves that rival those of Saudi Arabia, exhibits extensive ductile stretching as well (Fig. 105). Meanwhile, extending down the southeastern coast of the South American continent lie a string of large oil reserves that coincide with locations of ductile fracture arcs (Fig. 96). Into the corners of the massive North African ductile fracture, the two regions of extended ductile deformation are sites of two large reserves—one in Libya, the other extending from Tunisia into Algeria (Fig. 116). Finally, along the West African coast, a significant oil reserve lies directly on the ductile Bight of Biafra, where the borders of Cameroon and Algeria meet (Fig. 101). Perhaps some of the largest reservoirs remain untapped in Canada. Hudson Bay, the product of a very large ductile fracture, is a sedimentary basin which many believe has great potential for oil production.

This relationship between ductile deformation and major oil reserves (Fig. 119) suggests that the thinning and shifting of a continent's stratified layers may be directly responsible for the generation of oil—or at the very least, the release of oil. They may also make the oil more accessible from Earth's surface, as it lies within the shallow confines of a much thinner continental crust.

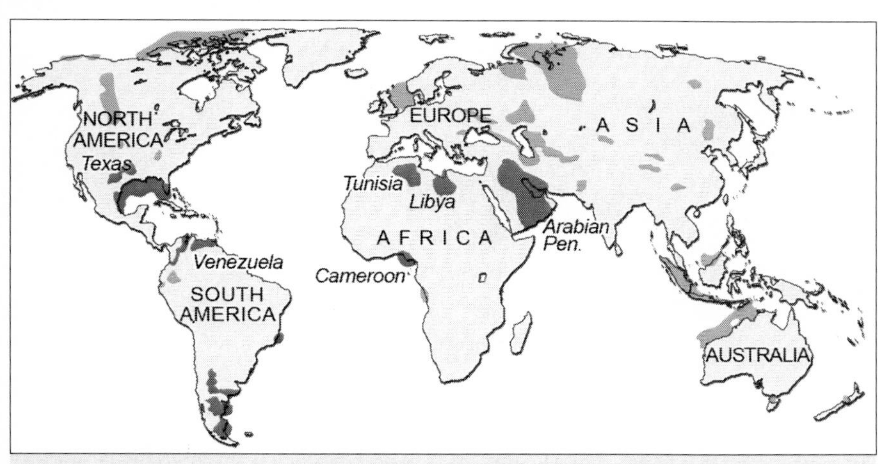

Figure 119. *Many of the world's largest oil reserves are situated precisely in the areas identified as locations of extreme ductile deformation.*

Moving on to the Indian Ocean, we find evidence beyond the Owen FZ linking East Africa to the Asian coast. Once again, formations in the Indian Ocean that have been identified as hotspot ridges are found to be boundary ridges. While the Chagos-Laccadive Ridge is currently believed to be the product of a volcanic hotspot named Réunion, its sister ridge the Mascarene Plateau is believed to be a combination of a hotspot ridge and a continental fragment. The new unified theory of boundary ridges recognizes again the mirroring of these two ridges across a central expansion ridge. Both ridges begin in the south at the Marie Celeste Fracture Zone and extend the same length and parallel up to the Vityaz Fracture Zone (VFZ). Beyond the VFZ, both ridges continue to mirror across the central ridge by veering up and away from each other (Fig. 120). The disparity in ridge lengths beyond the VFZ appears to be due to Madagascar breaking free of the African continent shortly after Africa broke free from Asia. The extra length can be found in ridge formations—the Amirante Bank and Farquhar Ridge—extending between the top of the Mascarene Plateau to the tip of Madagascar. Bringing the ridges back together mates the northern tip of Madagascar to the upper coast of India (Fig 120 A).

Major mirroring also occurs beyond these ridges, which links the southern tip of Madagascar to the Asian coast. The Kerguelen Plateau and Broken Ridge mirror the Southeast Indian Ridge in parallel, forming the lower extent of an expansion wedge. Similar to the previous ridges lying to the north, these ridges suddenly veer upward and away from the expansion ridge (Fig 120 B). Due to the movements of Antarctica and Australia and the opening up of the Southwest Indian Ridge, there are more complex transitions in these boundary ridges. Midway up the Ninetyeast Ridge there is a lone tangential rise, the Osborn Plateau, which records a sudden transition from ridges that mirror each other while rising away from the central expansion ridge to ridges paralleling it (Fig. 120 C-D). The top of this expansion wedge, where these ridges originated, reveals that the southern tip of Madagascar was initially merged with an area near Burma (Fig 120 D).

Taken altogether, Madagascar's northern point originally joined with India's northern coast (Fig 120 A), placing the eastern coast of Madagascar up along India's western coast. Meanwhile, India, as we

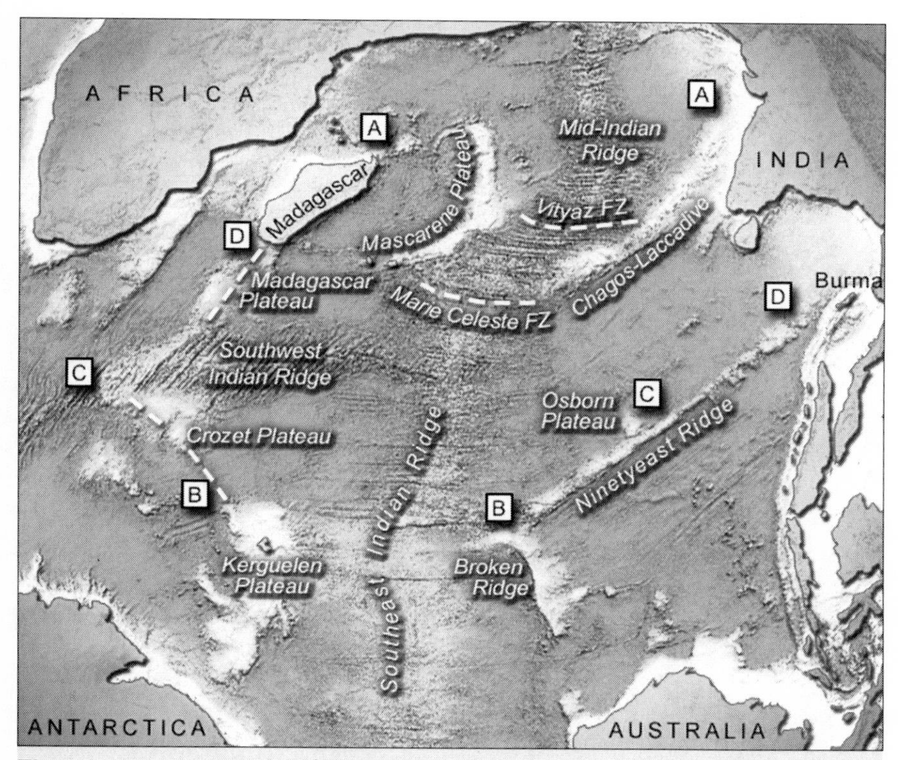

Figure 120. *Two instances of expansion wedges in the Indian Ocean currently viewed as a combination of hotspot ridges and continental fragments. One originating in the north is tied to Madagascar's northern fracture from Africa (A), and another originating in the south (D) is linked to Madagascar's southern fracture point from the continent.*

have seen demonstrated with other peninsulas, originally sat up against the coast of Asia before being pulled free with a clockwise-rotating Africa. This initial positioning of India allowed Africa and the southern tip of Madagascar to connect with the coast of Burma (Fig. 120 D).

Between the plate tectonics and Earth expansion theories, there is a significant difference in the interpretation of Africa's movement in relation to Asia. Based on the Owen Fracture Zone's link between the two continents, there is no denying that Africa was once attached to the Asian coast, but plate tectonics requires that the Asian coast was originally separate from the Asian continent. Geologists maintain that

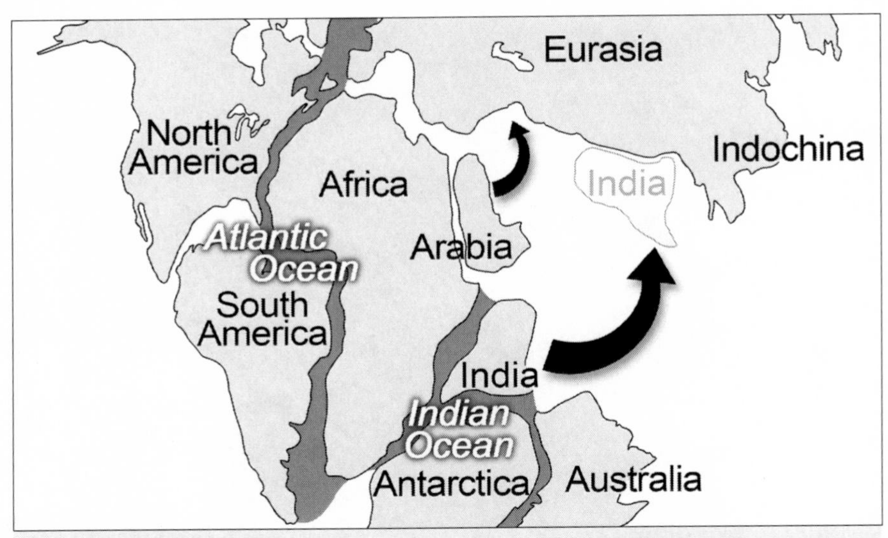

Figure 121. *While matching seafloor ages of the Atlantic and Indian Oceans seem to suggest that Africa was attached to the Americas and Asia simultaneously, the belief that Earth has always existed at its current size makes this an impossibility. Africa could have been joined to either the Americas or Asia, but not both simultaneously. Plate tectonics resolves this issue by positing that the southern coast of Asia was attached to Africa and migrated from Africa to Asia as the Indian and Atlantic Oceans opened up.*

a fragment of Africa separated and collided with Asia and now forms the southern coast of Asia. This interpretation is necessitated by the fact that the Indian Ocean and Atlantic opened up simultaneously. However, West Africa could not have been joined to the Americas at the same time East Africa was joined to South Asia with Earth at its current size. It would be similar to fitting a baseball cover over a basketball; bringing some portions of the cover together on the larger surface would require other portions to remain apart.

Geologists therefore decided that the Americas were initially merged with Africa. They posited a unique theory for India. In the widely accepted Pangaean model (Fig. 121), before the Atlantic and Indian Oceans opened up, India was merged with East Africa. Africa as a whole was oriented away from Asia and would remain in that alignment to this day. As the Atlantic and Indian oceans began to form and expand, India and other continental fragments broke free of African and migrated northward, forming the new southern coast of Asia. As evidence of this convergence, geologists cite the majestic Himalayas as a compression zone where two converging plates have caused extensive folding.

The theory initially places India's eastern coast up alongside Antarctica, which runs counter to new evidence placing India initially up alongside the Indochinese Peninsula. It also ignores the clockwise rotation of Qatar and matching surface patterns on the Arabian Peninsula, which imply that the whole of Africa rested along the Asian coast before it rotated 90 degrees to its current alignment, opening up the Indian Ocean behind it.

In order for Africa to lie up along the southern coast of Asia while simultaneously being joined with the Americas along its western coast, it would require the continents to exist on a much smaller globe. The evidence continues to build toward a much smaller Earth that has expanded over time.

We are still left with several questions—among them the true origins of the Himalayas—but our first goal is to determine how Earth could have transitioned from a planet with a unified crust to the heavily fractured planet we know today.

CHAPTER 12

GENESIS

Placing subduction and the plate tectonics model aside, we will now direct our focus toward the only remaining possibility: we live on an expanding earth. In this model, a unified singular crust initially encased a much smaller globe, one that was roughly 54% of its current diameter. This solid crustal shell eventually fractured and began fragmenting into continents, which drifted apart, while seafloor crust expanded to fill the vast voids forming between and the planet enlarged to accommodate the new and expanding surface area.

All Signs Point North

If we are to continue with our theory that the earth consisted of a single solid spherical layer of continental mass before fracturing and succumbing to expansion, it is critical to locate and study the area where the initial crustal breach occurred. Based on current radiometric dating methods, the likely location for this breach is just east of the Asian continent. It is here that we find the largest section of Earth's oldest seafloor crust, dating back 180-200 million years (Fig. 122).

It is no coincidence that this ancient seafloor sits near or alongside the linear path of peninsulas along Asia's coast. As we discovered earlier, peninsulas are fragmenting continental masses that are essentially

Figure 122. *The largest section of Earth's most ancient seafloor crust lies off the coast of Southeast Asia, making it the most probable location of the initial breach of Earth's once-unified crust. It is no coincidence that a series of downward facing peninsulas line the nearby Asian coast.*

splintering off from the mainland. They may be at different points in their fragmentation—Kamchatka and Korea remain attached to the mainland, while Japan has fully separated from Asia in the north—but they all share the same downward alignment. This alignment is no more than a curiosity among geologists who believe these peninsulas were formed randomly through three different means, but if all these instances are indeed fractured continental mass, the downward alignment of these continental splinters is very significant.

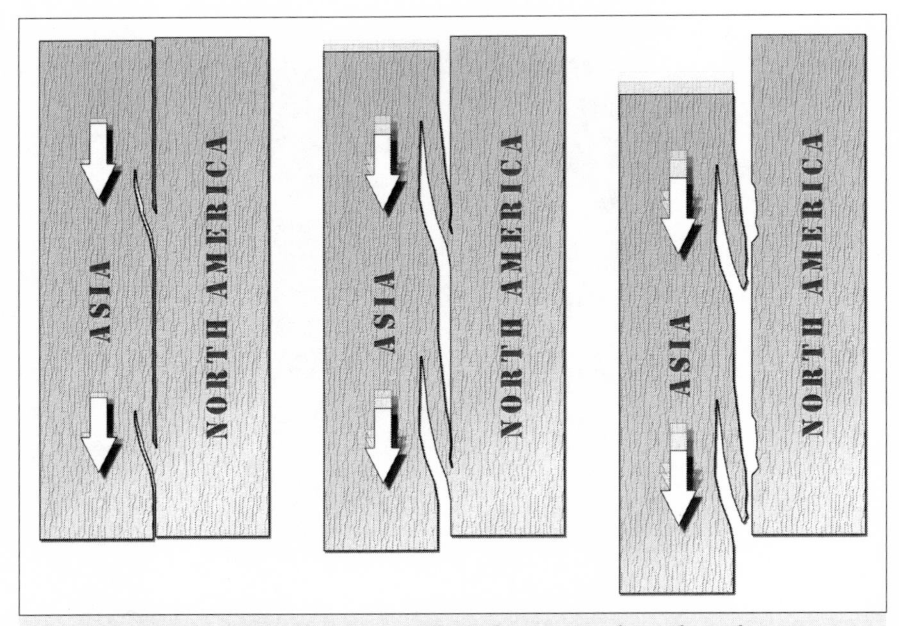

Figure 123. *Downward splintering replicated in an in-plane shear fracture occurring in a plank of wood. A similar dynamic likely occurred on Earth as Asia shifted past North America creating separation in the shearing process. This would prove to be the initial breach of Earth's unified crust and the birth of the Pacific seafloor in a region we also currently recognize as Earth's oldest seafloor crust.*

Splintering can be a sign that fracturing has occurred due to a sudden and violent impact from a specific direction and studying the splinters can potentially reveal the location of the impact. Consider the effects of an impact on a plank of wood. If we stood the plank on end on a solid surface so that it hung over the edge and we applied a forceful downward impact to the overhanging side of the plank in line with the grain, we would see the plank split with the grain (Fig. 123). While some of the grain would break away cleanly through the length of the plank, due to natural material inconsistencies, the grain would experience resistance in places, preventing a clean break.

In these instances of resistance, there is a transition point where the grain is partially separating from both halves of the plank yet firmly attached to both halves. (Fig. 123 center). As one side continues on its way downward past the other, the grain snaps cleanly from the piece experiencing less movement. This leaves downward-facing splinters lining the side of the plank that is moving downward (Fig. 123 right), in what is known as an in-plane shear fracture.

We can apply this same concept to crustal splintering. What began as a single intact plate enveloping an earth roughly half its current size was impacted from the north, creating a linear surface compression pattern extending down across the planet's surface. This compression can be seen on either side of the Pacific, where we find the coasts of both Asia and North America lined with mountain ranges forming linear compression zones (Fig. 124).

The impact compressed a path in the global shell, but for the fracture and subsequent splintering to occur along the Asian coast, the Asian region would have to be shifting past the North American region during the impact. The compression itself traveled down from the north to the then-smaller Earth's equator. Earth's crust may have remained intact throughout the impact were it not for softer crustal material in Asia. As the path of crustal compression halted at the equator, this softer material suddenly experienced major structural failure. The ductile crust began folding and compressing to an extent unrivaled elsewhere on the planet, creating the Himalayan Mountains, home of

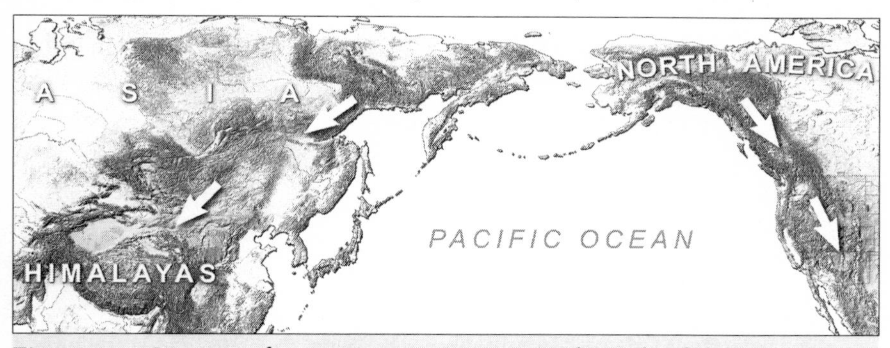

Figure 124. *Linear surface compression zones on either side of the Pacific Ocean along with the downward splintering along the Asian coast point to a terrestrial impact in the north. The shearing apart of the continents occurred when the Asian continent suddenly shifted past the North American continent. The cause of the shift? Softer ductile continental crust at the base of the Asian compression zone compressed more rapidly and dramatically forming the tallest mountain range on Earth, the Himalayan Mountains.*

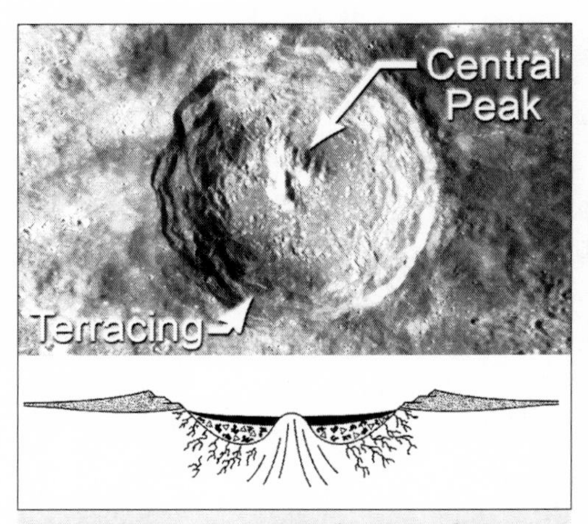

Figure 125. *Anatomy of a complex crater consisting of terracing lining the crater's inside perimeter and a central uplift, or central peak. (Top: NASA image of Eratosthenes lunar crater. Bottom: Cross-section illustration.)*

the highest peaks on Earth. It was this final extended crustal compression that allowed the Asian continent to suddenly and swiftly shift past North America in an in-plane shear. The final result would be the initial fragmentation of Earth's unified crustal shell, the formation of the downward splinter fractures in the form of the Asian peninsulas, and the birth of the Pacific seafloor.

This leaves us with the task of locating a major terrestrial impact site in the Arctic, above the Asian and North American linear compression zones. Considering that the crustal deformation we are referring to far exceeds any previously recognized terrestrial impact, logic dictates that we are looking for an impact that far and away dwarfs all known terrestrial impact craters—including South Africa's 185-mile-diameter Vredefort crater, currently recognized as Earth's largest terrestrial impact site.

If the impact crater remains visible on the planet's surface, there are key features that can aid in the identification process. When a celestial object such as a large meteorite, asteroid, or comet collides with a planet, it ejects terrestrial material, leaving behind a circular recess in Earth's crust known as a crater. Around the crater's perimeter, the outer edge proceeds to collapse downward, creating a terracing effect. Since this terracing occurs within the crustal recess of a circular crater, it lies in an arcing or circular pattern. Another potential identifying feature occurs if the crater happens to be a complex crater. Complex craters feature a central peak, which occurs as the center of the crater rebounds upward from the force of impact; it most often exists in the form of a centrally located consolidation of multiple peaks.

The Eratosthenes crater located on Earth's moon (Fig. 125)

provides an example of the central peak composed of multiple smaller peaks consolidated in the center. It also provides another visual that will potentially help in identifying the Arctic crater. Here we can see the circular terracing, which parallels the outer rim and extends nearly halfway between the rim and the base of the central peak. This terracing appears as uneven waves of material that clearly trend in a circular manner, running parallel to the crater's outer rim.

The Genesis Impact Crater

The importance of these key features—the circular terracing and multi-peaked central peak—lies in the fact that while the crater in the Arctic no longer exists intact, these unique elements remain. This is the site of the Genesis impact crater and it marks the beginning of Earth's transition from a single unified crust to a multi-fragmented crust.

In the center of the Arctic Ocean seafloor lies a geological feature that mimics crater terracing (Fig. 126). There is neither a crustal rim lining its outer edge nor does the terracing form a full circle, characteristics we would expect to find in a fully intact crater. There is no denying, however, that the terracing has an arcing trend that is fully discernable at each end of its semicircular form. Each end trends toward the Lomonosov Ridge, which marks the base of this semicircular formation.

If this is indeed the inner terracing of an impact crater, the abrupt edge at the Lomonosov Ridge might suggest that the crater separated along a radial fracture, which may be a shared characteristic with the Chicxulub crater. This is very significant, as the Lomonosov Ridge is one of two outer ridges paralleling the central Gakkel expansion ridge, and closing this expanded fracture back together brings about a remarkable alignment. Franz Josef Land, having all the characteristics of a multi-peak central peak, pockets directly into the center of the terracing. The other half of the crater's terracing lies beneath a continental shelf, which formed when sediment deposited up against the Barents-Kara Ridge also engulfed the crater's central peak, Franz Josef Land.

The crater's rim was formed by the surrounding continents, which

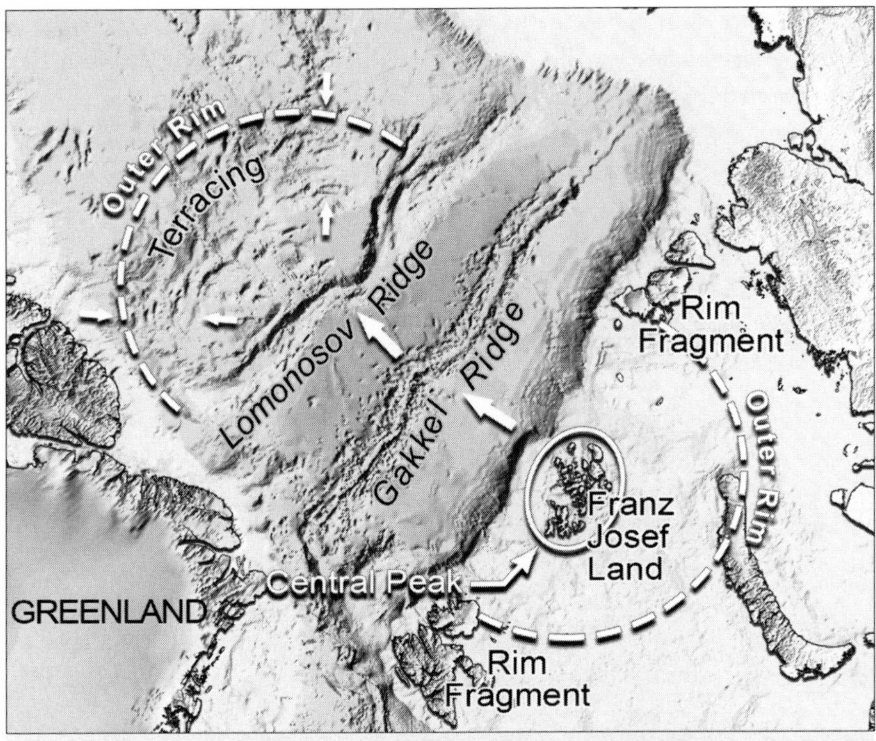

Figure 126. *A unique seafloor feature in the Arctic appears to mimic terracing found in a complex impact crater; the pattern's trending arc is key. Significantly, when the Gakkel expansion zone is closed back on its central ridge, which would be our potential crater's radial fracture, Franz Josef Land, sits at the center of the terraced arc, giving all the appearances of a complex crater's central peak. What appear to be rim fragments on either side of Franz Josef Land, along with the length of the terraced arc, reveal the original size of the crater to be 715 miles in diameter. This would make this impact crater over 3.5 times the size of the largest previously confirmed impact crater, South Africa's Vredefort.*

have since fragmented and pulled away from the crater's basin. Figure 127 closes down the Gakkel expansion zone and folds the Alaskan peninsula back up against the North Canadian coast, bringing the surrounding continental mass back into place. The result finds the central peak, Franz Josef Land, in the midst of the remaining void.

In the crater's current state, there remain rim fragments in island form to each side of Franz Josef Land. And as would be expected of a central peak, Franz Josef Land lies centered between these two rim fragments, which similarly lie along the Barents-Kara Ridge (Fig. 126). Based on the distance between these two fragments and the width of exposed terracing, the Genesis crater appears to have had a diameter of about 715 miles. This makes the Genesis crater the world's largest, over 3.5 times the size of Vredefort in South Africa.

Figure 127. *Franz Josef Land lies near the center of the Arctic void when the Gakkel and Atlantic expansion zones are closed back up and Alaska is folded back along the North American continent, suggesting that this island set is the central peak of an impact crater.*

Based on the 715-mile diameter, we can estimate the size of the celestial impactor as having been approximately 80-85 miles in diameter. Compare this to the Chicxulub impactor, which measured a mere 17 miles in diameter, and it is no surprise that the Genesis impact had a far greater effect on the planet's surface. As we have already seen, the impact was likely responsible for fracturing the planet's unified crust and initiating the formation of the Pacific Ocean, but there is also evidence that it affected far more than that. The effects are so staggering, in fact, that they leave little doubt that this Arctic feature is indeed the site of a major terrestrial impact.

In the immediate vicinity of the impact, the elongated island of Novaya Zemly and the Ural Mountains extending into the Russian mainland (Fig. 128) are currently believed to have been formed by the convergence of two ancient continental plates. It seems more likely, however, that the Ural Mountains mark the existence of a second radial fracture extending off the Genesis impact. The fracture's characteristics are very similar to those of the Cameroon line (Fig. 101, p. 199) and represent a fracture along the seafloor that extends deep into the Asian continent. The island of Novaya Zemly marks the seafloor portion of the fracture,

Figure 128. *Russia's Ural Mountains and the island of Novaya Zemly, believed to be formed by the collision of two continents, is a secondary radial fracture emanating from the Genesis impact crater. Like the Cameroon Line, this is a brittle fracture that has been subjected to stress allowing magma to seep through.*

while the Ural Mountains, mark the continental extension, both formed through magmatic upwelling. Running perpendicular to this formation are the Gakkel and Mid-Atlantic Ridges along with the Lena River, which also defines a radial fracture radiating out from the Genesis Impact.

The more extensive effects of the impact begin back at the original compression zone lining the Pacific, the mountains lying along the eastern coast of Asia, and all along the western coast of North America. The compression actually extends far beyond these coastal regions and is part of the much larger Genesis Hemispheric Impact Structure (GHIS), which originated from and remains centered on the Genesis Impact (Fig. 129). On the Asian continent, the Himalayas form the base of the Eurasian compression, but the compression extends linearly westward, encompassing the mountains of Afghanistan and Pakistan, the Zagros Mountains of Iran, the Caucasus, the Taurus mountain

Figure 129. *The Genesis Hemispheric Impact Structure, a dual layer of geographic rings originating in Northeast Asia and Alaska from the Genesis impact crater and encircling it. The main layer consists of a once-continuous compression zone that demarcates just under half of the world's landmass to the north and half to the south, revealing the two original hemispheric masses that formed the crust of a smaller Earth. Some continuity in these rings has been lost. Mexico's Sierra Madre have pulled out of alignment with North America's Appalachian Range due to the Tex-Mex ductile fracture arc (Fig. 112, p. 209) and the opening of the Atlantic separated the Appalachian Range from Morocco's Atlas Mountains—though as previously revealed, the Pico-East Azores Fracture Zone clearly links the two ranges before the continents fractured apart (Fig. 115, p. 213). The joined ring of mountains forms a downward compression that extends to and around the planet's ancient equator. The compression generated crustal displacement, partially shearing the compressed crust from the inner uncompressed land mass. This created the inner hemispheric rift composed of a parallel ring of the globe's largest and deepest inland bodies of water.*

range in Turkey, the Swiss Alps, the Pyrenees separating France and Spain, the Atlas Mountains lining the northern coast of Africa, and other smaller ranges. In North America, the compression seems to have continued southward to the southern tip of Mexico; as demonstrated earlier, Mexico and its Sierra Madre Mountains aligned with the Appalachian Mountains before the Gulf of Mexico opened up (Fig. 112, p. 209). Thus, similar to the Eurasian compression, the North American compression originally swept downward and away from the Pacific along the Sierra Madre and continued on to the Appalachians. More importantly, when the expansion of the Atlantic Ocean is reversed and North America reconnects with Africa, Africa's Atlas Mountains are revealed to be an extension of North America's Appalachian Mountains. The Pico-East Azores Fracture Zone confirms this original alignment (Fig. 115, p. 213).

We can see, then, that the entire pattern of compression originates from the Genesis impact crater in the Arctic. It extends down the side of the planet along the Pacific before diverting out and entirely around the globe, encircling the Genesis impact. The compression pattern's sudden tangential diversion from a linear downward pattern to an encircling ring would seem to suggest that the downward compression came up against a barrier. That barrier would most likely be the curvature of the earth on a planet half of Earth's current size. With the impact occurring near the northern pole, the equator represents an equidistant transition of Earth's curved surface.

Of note is the fact that the compression ring and land to the north comprise roughly 27.5 million square miles of continental mass. Adding the Genesis crater floor, which represents a displacement of continental mass, takes the combined original northern landmass to 28 million square miles. Meanwhile, all landmass to the south of the compression ring represents approximately 30 million square miles. Why is this significant?

The circular compression and land within it to the north comprises nearly half of the planet's continental landmass, and it may have been closer to half before the extensive compression and mountain building. In turn, refitting this northern impact structure onto a planet roughly 54 percent of its current size would see the outer compression ring sitting along this smaller planet's equator. It would appear, therefore, that on this smaller planet with a unified crust, the Genesis impact generated a hemispheric impact structure extending to or near its equator.

The impact's compression also generated another interesting hemispheric pattern. The compression zone reflects a great amount of crustal movement while inside the compression ring there is evidence of little to no crustal compression. We should expect to see effects similar to the initial breach of the Pacific. Shearing should occur as the moving, compressing portion of the continental mass shifts past the adjacent stationary mass—and, in fact, the shearing could not be more evident.

Immediately inside the compression ring lies a linear paralleling arc comprising the world's largest and deepest inland bodies of water (Fig. 129). We do not see total shearing between this outer ring and the land within, something we would expect if two rigid masses moved past each other. Instead, we find multiple instances of staggered shearing between the inner and outer masses, creating relief fractures stretching across both North America and Eurasia. These open fractures include the following inland bodies of water, which all have world rankings in the single digits in either area or depth (world rank in parentheses). Across Eurasia they are Lake Baikal (Area: seventh; Depth: first), Issyk Kul (Area: twenty-fourth; Depth: seventh), the Caspian Sea (Area: first; Depth: third), and the Black Sea, which is larger and deeper than any lake. Stretching across North America lie the Great Bear Lake (Area: eighth; Depth: thirtieth), Great Slave Lake (Area: ninth; Depth: eighth), and Lake Superior (Area: second; Depth: thirty-seventh).

The Mediterranean Sea is also included in this inner shear. In North America, the shear extends to and includes the separation of the Alaskan Peninsula from the northern mainland, while the Gulf of Saint Lawrence in the east is another instance of the shearing extending out to an open coastal fracture. Note once again the placement of the Pico-East Azores Fracture Zone, which reveals that the two separated shears or rifts existing on separate continents formed one continuous ring immediately after the Genesis impact occurred, perfectly paralleling the outer compression ring. The opening of the Atlantic Ocean would eventually break the continuity of both the compression ring and the inner shear ring, separating the Atlas Mountains from the Appalachians

Figure 130. *Currently, the age of most mountain ranges is based on or tied to their height. Shorter ranges are believed to be far older, having experienced hundreds of millions of years of erosion. Accepting the existence of the Genesis impact, we find that ranges like the Atlas and Appalachian Mountains are lower because they were formed on the back end of the impact, while the Himalayas and Rockies grew out of the direct linear force of the impact. More importantly, all these mountain ranges were formed within minutes of each other.*

and the Mediterranean Sea from the Gulf of Saint Lawrence.

After revealing these two extremely distinct paralleling patterns and their origins emanating from a potential impact site and encircling it all around, it is hard to imagine anyone still clinging to the belief that these features somehow formed and aligned through the random drifting and colliding of continental plates over hundreds of millions of years.

Orogeny, the science associated with the formation of mountains, currently maintains that taller mountain formations like the Himalayas (formed 45 million years ago, based on current radiometric dating methods) and the Rockies (55-80 million years ago) are more recent formations. Erosion over hundreds of millions of years transformed other ranges into low-lying mountains, like the assumed far older Appalachian (450 million years ago) and Atlas (300 million years ago) mountain ranges (Fig. 130).

Contrast this with a hemispheric impact structure. In this model, the Himalayan and Rocky Mountains reach their great heights as they absorb the direct force of the impact, the initial vertical path of surface compression. Meanwhile, the Atlas and Appalachian mountains, lying on the opposite side of the smaller planet's hemisphere, are much shorter

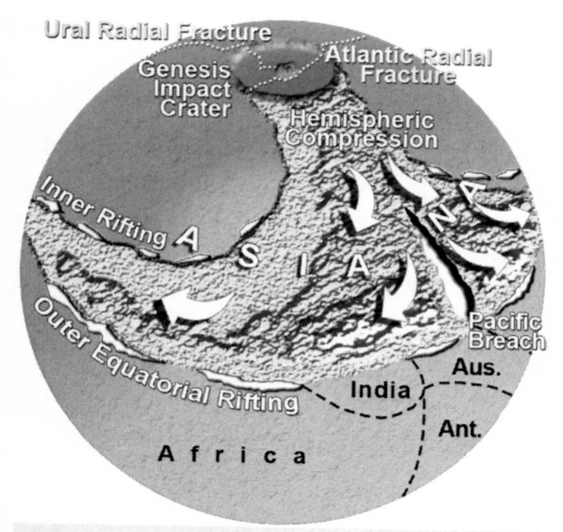

Figure 131. *Representation of the hemispheric impact structure immediately after impact and prior to continental separation and Earth expansion. The initial breach or rift between Asia and North America (NA) is also depicted.*

in height, having incurred only the less substantial residual compression effects of the impact. If this hemispheric pattern truly is the result of the Genesis impact, this ring of continuous mountain ranges came into existence within a span of minutes, not over a period of hundreds of millions of years.

Figure 131 approximates the hemispheric pattern laid over Earth at 54 percent of its current size. Here you can see the compression pattern extending vertically from the Genesis impact down to the smaller planet's equator. We see in the midst of the vertical compression zone the vertical shearing that occurred as the Himalayas compressed far more than the Rockies. This initial breach of Earth's unified crust would eventually open up into the Pacific Ocean. Midway down the planet, the compression redirects out along the equator and scales down significantly. This represents the subsequent formation of the lesser mountain ranges within the Genesis compression ring, like the Atlas and Appalachian ranges.

Inside the compression pattern, we can see rifting or shearing occurring as a result of the compressing and shifting of continental mass past the inner stationary mass. The fractures compose the northern hemispheric fracture ring, which makes up the largest and deepest inland bodies of water in the northern hemisphere, including the Mediterranean Sea, all lying in a ring extending down and around the Genesis impact crater.

An equatorial fracture ring was also created at the time of impact.

Due to shifting and deep folds occurring along the Equator, significant rifting and shearing also occurred along the smaller planet's equator, similar to that occurring above the compression zone. The result was the nearly complete shearing away and separation of continental mass to the south, with Australia and Antarctica eventually becoming completely independent of other landmasses. Africa and Arabia barely remain attached to Asia by an 800-mile swath of land, and South America, sheared from the Sierra Madre, dangles by a thread of an isthmus from North America.

The concentric patterns of compression and shearing emanating from the potential site of the 715-mile diameter Genesis impact crater and arcing across the two northern continents appear to confirm a hemispheric impact event. They also seem to be tied to the initial fracture of Earth's once-unified crust, the moment Earth expansion began.

We have now established several aspects of the Earth expansion theory. We have identified a hemispheric impact structure, a pattern of compression and rifting that encompasses nearly half of Earth's continental crust, supporting a smaller hemisphere on an ancient Earth roughly half of its current size. We have also demonstrated a consistent unified global pattern linking seafloor ridges with continental fractures which, when recognizing the Hawaiian-Emperor seamount chain's alignment with the cusp of a ductile fracture, refutes subduction, the most essential element of the plate tectonics theory. Further, we have demonstrated a consistent pattern of continental separation and negated all instances of significant convergence, which supports the theory of Earth expansion and runs counter to the chaotic plate movement that characterizes plate tectonics.

Now we must finally answer the difficult questions we have left unaddressed. We need to determine how a much smaller Earth, after the fracture of its unified shell, transformed to the much larger current-day Earth, with fragments of continental crust spread about and separated by large oceanic basins. And, of course, we must determine what dynamic could potentially drive a planet to expand in such a manner while also providing a means of stabilization in growth in order to conform to a modern-day Earth, which exhibits little to no sign of expansion.

CHAPTER 13

COPHEE: A NEW DYNAMIC

There is an enormous challenge in determining not only the mechanism that drives expansion but also the mechanism that switches it off. Yet like many of our previous realizations, the answer may have been staring us in the face all along.

Some have predicted that significant portions of Earth will eventually be overrun by the world's oceans, a popular view that spawned a blockbuster movie portraying a post-apocalyptic world where survivors dwelled on makeshift floating cities. A water world is an intriguing concept, considering we happen to live on a rather unique planet that not only possesses water but possesses it in extraordinary abundance—it covers 71 percent of the planet's surface. While we probe the solar system for scant traces of water, our planet bathes in it.

Let us consider how these waters would have impacted an Earth 54 percent of its current size encased in a unified shell, a shell consisting of all the world's continents and continental fragments merged back together around a much smaller Earth. This would have been a unified shell devoid of both oceanic basins and significant mountain ranges, yet to be formed by the Genesis impact. With no basins to contain the waters that make up today's oceans, these waters would have existed as a single ocean enveloping the entire planet, submerging all land deep beneath and making ancient Earth a true ocean planet.

Earth contains 332.5 million cubic miles of water, with our oceans averaging 12,100 feet (2.3 miles) of depth. This same volume of water sitting upon the unified crust of the smaller Earth would have sat at 29,000 feet (5.5 miles) deep prior to the Genesis impact. At 12,000 feet in depth, water exerts over 5,300 pounds per square inch (psi) of pressure on the planet. A 29,000-foot ocean depth would have exerted a global force of nearly 13,000 psi, more than double today's average.

Man-made reservoirs hundreds of feet in depth exert pressure on Earth's crust and have been known to generate seismic activity after their creation. While most of this seismic activity has been of relatively low magnitude, it is believed that some reservoirs have been responsible for quakes reaching up to 6.0 magnitude. These include a 1967 quake associated with the Koyna Dam reservoir in India and a 1975 quake associated with man-made Lake Oroville in California. The quakes appear to occur as water introduced into the reservoirs applies pressure on existing faults.

Now imagine a half-sized Earth. Instead of a few hundred feet of standing water generating nearly 100 psi in a reservoir, a global ocean 5.5 miles deep is generating 13,000 psi of force across its entire surface. As long as the crust remains intact, the weight distributes evenly; the crust, while under extreme duress, remains stable. However, what happens when a terrestrial impact suddenly ruptures and fractures that crust?

The Genesis impact generated a catastrophic effect on the planet's crust, forming the hemispheric compression zone, which would prove to be only the beginning of Earth's initial deformation. When the impact sheared Asia from North America, the overlying waters surged violently into the freshly formed breach. With the initial surge being limited to the Pacific breach, the effect on global sea level would have been relatively negligible at first. But the rapidly increasing water depth extending downward into the initial breach would have exerted unimaginable penetrative force. While the waters remained at 5.5 miles deep elsewhere, the breach in the continental crust allowed this 5.5 miles of water to crash violently downward through 20 miles or more of fractured crust. The cumulative 25.5-mile column of water would generate an incredible 60,000 psi.

This catastrophic surge of pressure began prying and splitting the Pacific breach open further and further, forming the most ancient portions of the Pacific seafloor. The force and increased momentum of this initial seafloor spread began weakening the crust and opening more fractures. The Atlantic fracture, an extension of the Arctic Gakkel radial fracture formed by the Genesis impact, would give way in the south to form the Atlantic seafloor, followed closely by the Indian Basin.

At this point, the planet began to expand to accommodate the new and rapidly growing seafloor. Of course, during this process of expansion Earth's core would have been highly active, expending an immense amount of energy. And in fact, it seems that this period of heightened core activity did not go unrecorded. Magnetic striping spread across the oceanic basins indicates a series of pole reversals having occurred during seafloor creation. The current assumption, based on seafloor crust believed to be 260 million years old, is that these patterns record sporadic random pole reversals that occur anywhere from 10,000 to 25 million years apart. It is believed that instabilities in Earth's geodynamic core, which currently cause the magnetic poles to meander incrementally about the geographic poles, periodically become far more unstable, producing a complete shift in the magnetic poles.

The catastrophic nature of the Genesis impact and subsequent surge of waters dictate a very rapid rate of seafloor creation. Seafloor crust was not being generated due to the slow and steady movement of invisible subterranean forces; this was a catastrophic surge of waters violently pressing down on and through the continental crust, carving up the planet while seeking a new resting place on the hot but cooling magma lying deep below. Therefore, the expansion process likely spanned a relatively short period of time—days, rather than hundreds of millions of years, as currently held. Like erroneous dating of the major mountain ranges, current methods of dating the seafloor now also appear flawed.

This shortened timeframe for seafloor creation means that the pole reversals recorded a period of heightened activity in Earth's core. Earth's core reacted to the sudden crustal compromise and became highly active, spinning at a much higher speed than normal and generating intense energy directly linked to the planetary expansion process. The accelerated spin generated pole reversals that were occurring at a rate far more frequent than thousands or millions of years apart. Based on a more rapid seafloor creation process, reversals were cycling through over the course of days.

Recent research led by Peter Sturrock of Stanford and Jere Jenkins and Ephraim Fischbach of Purdue University concluded that the rate of radioactive decay fluctuates and can accelerate, an effect directly linked to the rotation of the sun's core.[71] It is believed that the effect is currently minimal due to the extreme distance between the sun and Earth, but what effect might we expect if similar activity occurred much closer to Earth—or, for that matter, at Earth's very core? The increased activity of Earth's core recorded in a pattern of seafloor magnetic striping may potentially have rendered radiometric dating methods completely unreliable.

There is more evidence supporting rapid seafloor creation and, by extension, rapid planetary expansion. This lies in two very distinct periods of expansion marked out in a consistent global pattern. Mirrored ridges throughout the globe extending out from the continents abruptly end and remain separated by 30 to 50 million years of expansion crust, based on current dating methods. We previously examined two expansion wedges in the Indian Ocean (Fig. 120), neither of which remain fully intact. At the base of the larger wedge we find what were once joined ridges, Ninetyeast and the Crozet Plateau, separated by 1,200 miles of seafloor crust believed to have formed within the last 30-50 million years (Fig. 132). This same crust and central expansion ridge also separates the Kerguelen Plateau from Broken Ridge, its mating formation, and the Mascarene Ridge from its sister ridge, the Chagos-Laccadive Ridge. Meanwhile, a narrower span of 30-50 million-year crust separates the Madagascar Plateau from the Crozet Plateau.

In the Atlantic, the Walvis Ridge and Rio Grande Rise extend off the continents of Africa and South America respectively. Like the mirrored ridges of the Indian Ocean, we find the two mirrored features stopping

---

[71] P. A. Sturrock, J. B. Buncher, E. Fischbach, J. T. Gruenwald, D. JavorsekII, J. H. Jenkins, R. H. Lee, J. J. Mattes, and J. R. Newport, *Power Spectrum Analysis of BNL Decay Rate Data, 2010*

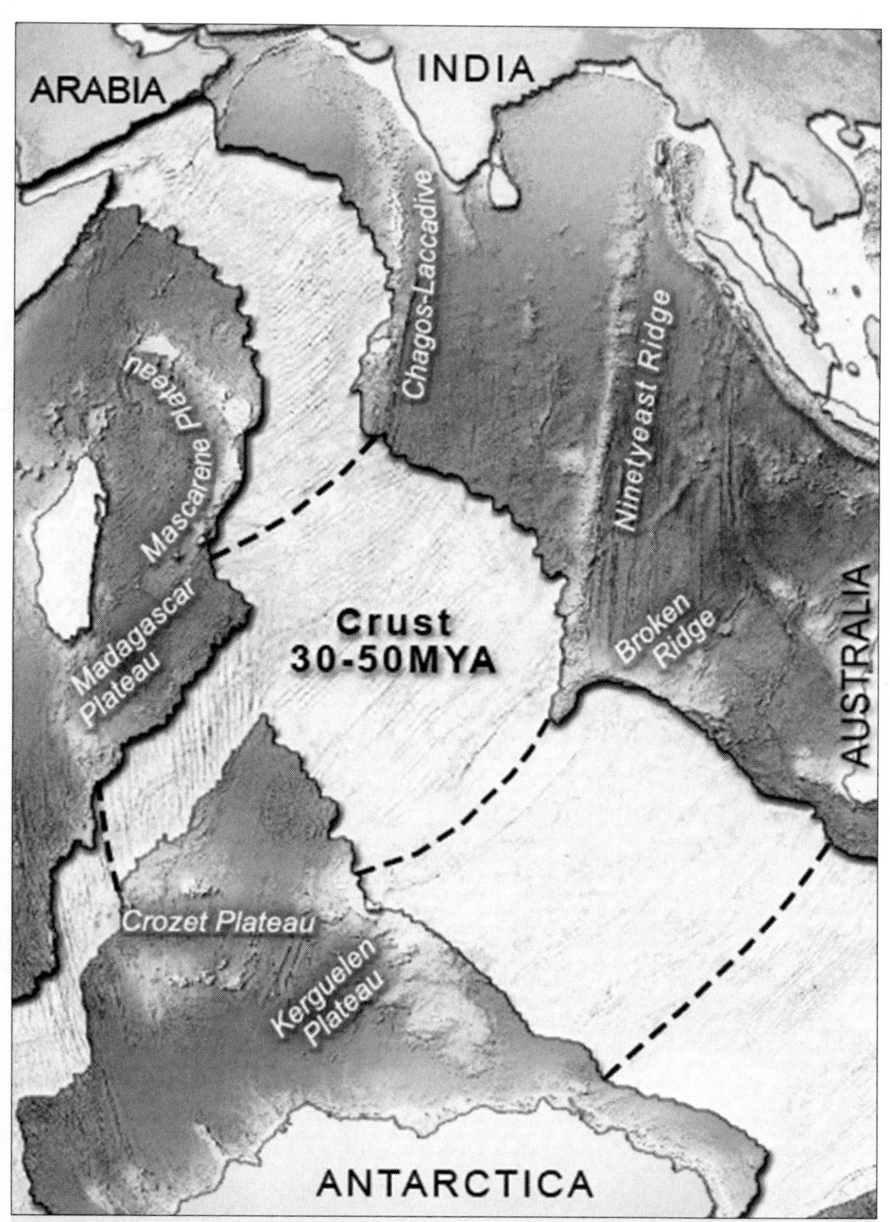

Figure 132. *In the Indian Ocean and throughout the globe, mirrored seafloor ridges consistently extend to seafloor crust currently dated to roughly 30-50 million years old and abruptly stop. Ridge formations that were once joined at their furthest extremity away from land were fractured apart during a second expansion event.*

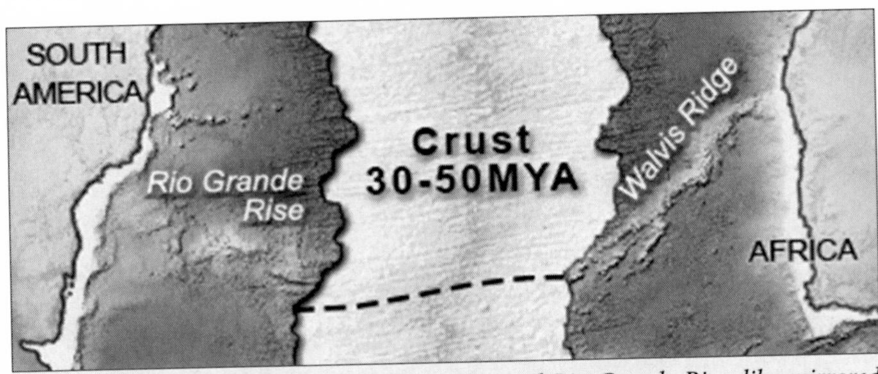

Figure 133. *In the Atlantic the Walvis Ridge and Rio Grande Rise, like mirrored ridges in the Indian Ocean, are separated by seafloor crust currently dated between 30-50 million years old that has experienced little to no extended ridge formation.*

short of the newer swath of seafloor crust believed to be 30-50 million years of age (Fig. 133).

This consistent pattern reveals that, based on the 30-50 million year dating, all of Earth's expansion wedge boundaries were joined to form downward-pointing V-shaped ridges before simultaneously separating at their lowest point with little to no extension beyond. This is a globally consistent pattern, with the exception of one set of ridges.

The Hawaiian-Emperor seamount chain and the Baja Peninsula are separated by 3,000 miles of seafloor crust. This crust is currently believed to be twice the age of these other crustal spans, with dating ranging from 50-100 million years (Fig. 134). This means that based on current dating, 30-50 million years in the past, while all other ridges extending out into their corresponding seafloor were joined in V-shaped formations, the Baja Peninsula and Hawaiian-Emperor seamount chain were the planet's lone detached ridge set. Fifteen hundred miles of seafloor crust separated the two ridges, which is roughly half of the current distance. This, of course, assumes that the current span of crust was generated along the North American coast and drifted westward, leaving new crust lying adjacent to the Baja Peninsula and older crust lying adjacent to the Hawaiian Islands.

However, the existence of the North Pacific divergent boundary centered between these two ridges dictates that the newer crust lies centered between the Baja Peninsula and the Hawaiian Islands at the

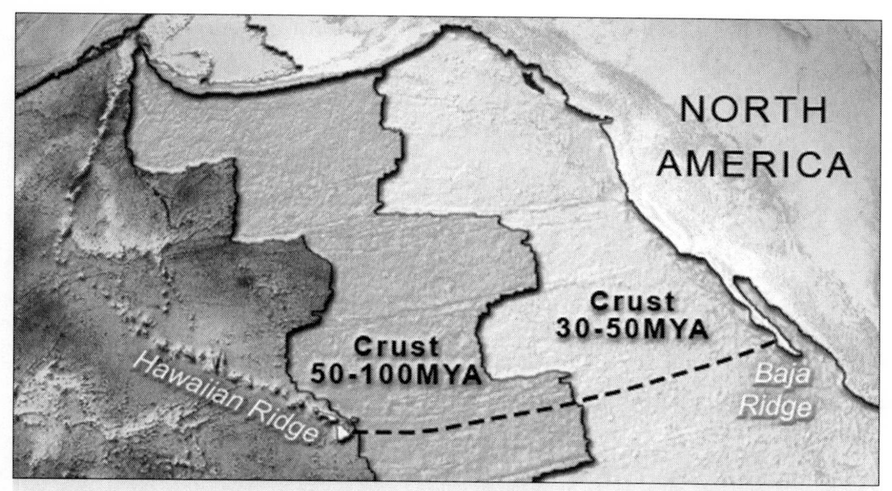

Figure 134. *The Hawaiian Ridge is separated from its sister ridge, the Baja Peninsula, by seafloor crust believed roughly twice the age of the seafloor crust that separates mirrored ridges in the Atlantic and Indian Oceans. With the strong possibility of a divergent boundary resting between these ridges, it is likely that this span is the same age as all other spans between ridges. Instead of the increased age moving east to west as shown, newer crust has formed in the center and spread outward. In other words, both halves of the span are identical in age range.*

site of the divergent boundary, and the older crust lies adjacent to either side of the two separated Hawaiian and Baja Ridges, similar to the other mirrored ridges throughout the globe that form expansion wedge boundaries. This mirrored pattern of crustal aging extending from the center outward, of course, would mean that the current east-to-west aging is erroneous and does not allow for the gradual aging between these ridges that is believed to extend to 100 million years. We have already exposed the potential flaws in radiometric dating as it pertains to mountain building in the Genesis impact structure; it logically follows that seafloor crustal dates should be viewed with similar skepticism.

With the North Pacific divergent boundary indicating a significantly reduced age for the span between the Baja Peninsula and the Hawaiian-Emperor seamount chain, I am convinced that the span of crust lying between is similar in age to the other crustal spans lying between ridges in the Atlantic and Indian oceans. Therefore, in all likelihood, all expansion wedge ridges throughout the globe were joined and became separated simultaneously in the past. The point at which all of these ridges were at their greatest extension but still fully joined marks the date of the first expansion event. The section of new seafloor crust that currently dates from 30-50 million years, including the span between Hawaii and Baja, forms a continuous band extending out and across all oceans, reflecting a period of secondary expansion shared across all ocean floors. This band consistently forms the span between mirrored ridges, which abruptly cease formation along each side.

Having established that these ridges are tied to continental fractures and exist as boundaries for expansion wedges, throughout the globe these wedges were fully formed, well-defined ridges consistently joined in the south forming downward Vs. These boundary ridges were created as lava flowed through fractures in the seafloor. This flow was facilitated by continuous movement and shifting between the seafloor plates on either side of these boundary fractures.

If the bordering plates had been moving at the current near-static speed of inches per year, it is easy to imagine these boundary fractures would bond immediately on conception. But the ridges extend hundreds or thousands of miles out from their point of origin. This means that seafloor plate movement and expansion was a continuous process until suddenly these magma-seeping fractures were solidly bonded throughout the globe, preventing the fractures from extending into the newer band of seafloor crust.

Plate tectonics conforms to the uniformitarian view that plates have moved at the same relative speed throughout time, the rate of a mere few inches per year. But if the fractures were indeed able to extend out from the continents to such great lengths at that slow and steady rate, there is no clear reason for the ridges to suddenly halt formation. Perhaps it is reasonable to accept that one of these V-shaped wedges could form, stop formation, and then slowly drift farther and farther apart as the new crust continued to form between with no ridge extension. But there is little to no explanation for how this could happen to every V-shaped wedge simultaneously throughout the globe based on the slow seafloor expansion proposed by plate tectonics.

In keeping with the catastrophic nature of the Genesis impact and the cataclysmic surge of overlying waters into the initial breach that followed, it is far more probable that the seafloor was expanding at a rapid catastrophic rate—likely miles per day—before coming to a relative stop. This first expansion event was followed by a long period of global stability between all plates. This stability allowed the magma seeping up through the plate boundaries during and immediately following the expansion event to solidify. Thus, when the Earth experienced what clearly appears to be a second expansion event, the now fully bonded expansion ridges extended no farther into the newly generated seafloor crust.

Globally, this second expansion differentiates itself not only by this sudden discontinuation of ridge formations, but also by newer crust that exhibits far fewer blemishes. Nowhere are seafloor blemishes more abundant than where the waters first violently breached Earth's crust. There is a huge contrast between the West Pacific seafloor basin and all other basins. The initial mass surge of waters into the West Pacific breach left the seafloor scarred with an inordinate amount of volcanic features. The Pacific contains the world's highest concentration of atolls; millions of square miles of seafloor are littered with these islands, which have formed over extinct volcanoes. The Mid Pacific Mountains, a vast undersea range lying just east of Wake Island and extending to Necker Island on the lower end of the Hawaiian-Emperor seamount chain, has all the appearance of a writhing beast casting volcanic ridges off its body in a northeasterly direction. This location for the initial breach contributed to making the Pacific Earth's largest and deepest ocean, and the seafloor bears the scars from the extreme weight of the waters violently heaved upon it.

The initial breach also likely contributed to another unique aspect of the Pacific. Unlike other downward-pointing V-ridges, one side of the Pacific's V-ridge resides almost entirely up against a continent and above sea level. This difference is likely due to the effects of the distinctly different fracture dynamic occurring in the Pacific breach. All other continental fractures are opening-mode fractures. In other words, stress put on the continental mass in these regions forced them to split and separate along the same plane away from the central fracture. Think of it as simply sliding two objects on a flat surface away from each other.

By contrast, the Pacific basin was generated by a shear fracture, the

Asian continent having slid past the North American continent. The impact of the shear fracture in the Pacific is extremely significant in that it occurs across a curved surface. What began as an in-plane shear as Asia shifted past North America quickly transitioned into an out-of-plane shear. The Asian and North American coasts shared points above and below throughout the shearing process, but clearly if one side of the shear were to have become shorter than the other, it would not be able to arc as high as the longer side of the shear. The compressed and shortened path to the Himalayas along the coast of Asia could not remain on the same plane as the less-compressed, longer stretch of the North American coast, which resulted in the out-of-plane shear. This left North America sitting on a higher plane than Asia after the shear.

The other oceanic basins, which experienced opening fractures, saw the seafloor expand equally on each side of a central expansion ridge. This also resulted in fractures or blemishes on the seafloor retaining symmetry across the expansion zone.

The out-of-plane breach in the Pacific, however, created an imbalance. As the waters surged into the initial breach, we not only witness the occurrence of violent upheaval along the seafloor, leaving the Western Pacific heavily scarred, but we also find the two sides of the fracture searching for equilibrium—two continental plates seeking to return back to the same plane.

Gravity favoring the Asian continent sitting at the lower plane may have seen seafloor more easily falling away from the North Pacific divergent boundary toward the lower-lying continent. The Hawaiian-Emperor seamount chain therefore extends far out across a significant portion of the Pacific seafloor, while the Baja Peninsula and the rest of the North American Coastal Ridge hug the North American coastline. This imbalance also allowed the NACR to remain at the higher sea level until equilibrium was reached at the end of the first expansion event. By this time Asia had finally risen above sea level to sit along the same plane as North America, and the Hawaiian Islands, nestled in the side of the Baja Peninsula just off the North American coast, also sat above sea level.

After the Earth attained equilibrium following the first event, the second expansion event saw more balanced expansion in the Pacific, similar to what had occurred previously in the Atlantic and Indian oceans. Patterns of mirroring began to occur in the newer crust of the eastern Pacific across the North Pacific divergent boundary, with features sitting nearly equidistant to either side.

But what brought about two subsequent periods of relative dormancy—the period following the Genesis impact and the current period of relative inactivity that has followed the second expansion event? The halting of the expansion process has been perhaps the biggest obstacle for Earth expansion theorists. The logical expectation is that once expansion begins, the increased energy generated within the planet would start a powerful chain reaction that would continue until the planet either exploded or collapsed back in upon itself. If Earth has truly experienced planetary expansion in the past, what dynamic could possibly halt the process so that we arrive at the point today when little to no expansion is being recorded? It would appear that the same element that drives expansion is also tied to the dynamic that abruptly halts the process: water.

As mentioned earlier, Earth is unique in its abundance of water. This fact has raised questions as to the origins of the planet's water and why Earth possesses an abundance not seen elsewhere. Yet perhaps we have neglected to ask another seemingly relevant question.

Imagine if we were to fill a warehouse with hundreds of open and empty containers and one day found that one lone vessel contained a few drops of water. We'd be left with a simple question: Where did this water originate? Now imagine a second scenario where that same lone container is full to the brim and water lies all around it. Again, where did it come from? Now imagine a third scenario where that container is full to the brim with water but not a trace of water can be found beyond the container on the surrounding floor. The initial question would still arise, but curiosity would drive a second question: How did it happen that the water stopped filling the glass precisely as the glass was at its fullest?

Earth poses a similar mystery that does not seem to draw much attention. Imagine Earth's oceanic basins devoid of all water. In Earth's current form, the continents would be separated by immensely large arid basins. These deep canyons would compose two-thirds of Earth's surface and leave the world with a unique otherworldly landscape, its continents existing essentially as large isolated mesas. It would be a rugged planet nearly impossible to traverse from one continent to another. Traveling internationally would involve an experience much like climbing down one side of the Grand Canyon and up the other, only these canyons would be more than twice as deep and thousands of miles wider.

In this hypothetical scenario, these canyons exist as one extremely large open container. Like those in the warehouse example, if you found there to be a few small seas or lakes sitting at the very bottom of this expansive basin, you might simply wonder about the water's origins. If, on the other hand, this vast basin were not only completely filled to the brim but beyond, water having washed over much of the continental masses so as to leave little to no land rising above its surface, again you'd have the same question.

But that is the extremely intriguing aspect of our planet's relationship with water. Earth does not exist in either of the two hypothetical states. Instead, these enormously deep and vast basins that meander across two-thirds of our planet are filled right to the brim with water. There is no place on Earth where these waters overrun continental crust.

Perhaps more to the point, science recognizes two features that are currently unique to our planet. One is water on the planet's surface and the other is the dual nature of Earth's crust with the existence of thin low-lying crustal plates which have expanded to cover an astonishing two-thirds of our planet. Both of these features have origins science is currently unable to explain. Yet, somehow, these two extremely unique and seemingly unrelated features not only exist on one planet but also precisely occupy the same space on that planet allowing the third surface element, continental plates, to sit above vast oceans.

On Earth these two celestially unique features perfectly offset each other creating a delicate balance between dry land continents and oceans that we currently take for granted. If planetary expansion is a reality, there must exist a dynamic responsible for the sudden cessation of expansion and I believe this extraordinary balance of land and sea could be a key player in a highly complex planetary dynamic.

Catastrophic Ocean Planet Hydro-Equilibrial Expansion

We will be referring to this newly proposed dynamic as Catastrophic Ocean Planet Hydro-Equilibrial Expansion, or COPHEE. It posits that there is a uniqueness to ocean planets like Earth that potentially comprises three states.

The first and original state finds the planet in a static stable state of non-growth, possessing a unified crust enveloped entirely in water. An ocean planet like Earth is prone to undergo a dynamic transformation when the planet's crust is catastrophically compromised. Given sufficient depth and force from the overlying layer of water, the planet enters the second state, its expansion state, with fractured continental mass pried apart opening deep basins that expand to accommodate the sudden surge of waters. During this process, Earth's core becomes highly active, spinning at an accelerated rate and expending an enormous amount of energy. The core's instability during this period drives a continuous cycle of pole reversals.

This state of expansion appears to continue until the third and final state occurs: planetary equilibrium. Equilibrium occurs when the waters completely subside from the continents, draining into the oceanic basins, which expand in size to contain the continental runoff. If true, this is an extremely complex dynamic that finds Earth's deep interior core reacting to a relatively thin outer surface area, spinning at an accelerated rate throughout the expansion process and returning to its normal rate of spin once this thin outer layer achieves equilibrium.

Similar to a shaken container of oil and water where the fluids remain active until equilibrium is attained with the complete separation of oil from water, the Earth appears to enter a highly volatile period of expansion that continues until the oceans recede completely from the continents. The consistency of continental mass to maintain itself above sea level regardless of its thickness seems to support this. It is understandable that the more mountainous regions exist above sea level, this being the thickest of Earth's continental crust, but more surprising is the state of the planet's thinnest crust.

Ductile fractures form the planet's thinnest continental crust. These are regions that have been stretched and thinned to their limits, but in each instance these arcing spans remain fully intact above sea level. Lengthy spans like that which line the Gulf of Mexico, the many ductile arcs cut into the southeastern coast of South America, and the shores of the Hudson Bay all lie above sea level, with no sections overrun by the seas that wash upon their shores.

With society's current presumption that this balance could easily be upset with the melting of the Antarctic ice cap, we should question how we came to exist precisely when this delicate balance has occurred. Is it merely coincidence that finds us residing at the very edge of this apparent precipice? It seems unlikely considering the infinite range of possible sea levels—from a few small seas lying at the bottom of these deep, vast oceanic basins to continents completely overrun with water.

If COPHEE builds any sort of consensus, it will take time to formulate a more detailed understanding of such a complex dynamic. Yet based on evidence provided thus far that planetary expansion does occur, that oceanic basins are the product of a surge of overlying waters into a planet's fractured crust, and that Earth's vast oceanic basins are filled precisely to the brim with water, this hypothesis definitely merits consideration.

CHAPTER 14

COPHEE II: THE GREAT FLOOD

By the end of the first COPHEE event, the Earth had expanded a great deal, but it still remained substantially smaller than the planet we live on today. In addition, the continental crust, though heavily fractured and spread out across the globe, remained nearly completely joined as one large super-continent. We surmise this by taking away the planet's swath of newer crust, which is currently believed to be 30 to 50 million years old; this comprises the crust currently separating all of Earth's broken V-shaped expansion wedges.

Following the Genesis event, land-dwelling creatures of all sorts were able to migrate across all continents, from the southernmost tip of Africa to Europe and Asia. Asia and North America remained attached, as Alaska was in the midst of sliding down and out of the Arctic Ocean, and even Europe was likely linked to North America via Greenland. North and South America were more connected than they are today, with more of Central America laid up against South America. South America was in turn attached to Antarctica in the south; Australia's entire southern coast also lay against Antarctica. It is also possible that the merged Australian-Antarctic landmass remained linked to Indochina via Sumatra and Java.

Earth remained at this intermediate size for an indefinite period of time. Though the length of this period is likely impossible to determine due to potential flaws in radiometric dating, it would have allowed sufficient time for lava that had seeped between plates during the planet's unstable growth to cool and harden, bonding expansion wedge ridges all across the globe. Fully formed V-shaped scars marked the floors of the Pacific, Atlantic, and Indian oceans at this time. With the planet's crust fully bonded, it once again became stable, much as it was prior to the catastrophic crustal breach.

During this period, the climate also stabilized, warming the planet into a global paradise. Even the Polar Regions were allowed to thaw and warm to tropical temperatures. This is confirmed by the presence of tropical vegetation discovered lying beneath the ice shelves covering both Greenland in the north and Antarctica to the south.

Then a new expansion event occurred, which brought about the world we see today. At its current size and state, many continents have become completely detached, and a cooler climate has caused icecaps to form over vast polar regions. This second expansion was characterized by the splitting and separation of the globe's oceanic V-ridges, which had previously solidified and stabilized the seafloor crust. The central divergent boundary that extended through the center of each wedge gave way to new seafloor expansion, leaving the once fully formed V-shaped wedges separated by vast stretches of new seafloor basin between. This new seafloor is devoid of the catastrophic blemishes, including the expansion wedge ridges; today we only find them in the older crust. This global phenomenon of a newer span of seafloor lying between broken V-shaped scars supports the occurrence of a second major COPHEE event, COPHEE II.

However, we have a problem. How do we identify the source of the water that would eventually fill this expanded seafloor basin? In order to expand to Earth's current size, there would had to have been a sudden influx of water so great in volume that it would fill three-quarters of today's Pacific Ocean basin. While comets are made partially of ice, the volume of water delivered by a comet strike would be rather insignificant—just enough to form a small body of water at best.

The answer to COPHEE II's water source may lie within an ancient text. If the story of Earth experiencing two major COPHEE events sounds familiar, it is likely because it mimics the Biblical accounts of creation and the Noachian Flood. According to the Bible, Earth began as an ocean planet, and on the second of six creative days, God caused a separation of land and sea:

> *In the beginning God created the heavens and the earth. Now the earth was formless and empty, darkness was over the surface of the deep, and the Spirit of God was hovering over the waters.*
>
> *And God said, "Let the water under the sky be gathered to one place, and let dry ground appear." And it was so. God called the dry ground "land," and the gathered waters he called "seas." And God saw that it was good.*
>
> *And there was evening, and there was morning—the third day.*[72]

The Bible's account of the first creative day coincides with the Genesis terrestrial impact and the subsequent COPHEE dynamic—COPHEE I—which involves the complete separation of water and land.

There exists another biblical scripture pertaining to this separation of land and water that is equally intriguing:

> *When he gave the sea its boundary so the waters would not overstep his command, and when he marked out the foundations of the earth.*[73]

While this passage could easily be seen as a man's simple observation ascribing credit to God, it seems to suggest more—the acknowledgment of a mechanism that ensures that the oceans cannot overrun the continents. This, of course, is the very essence of COPHEE.

As an aside, it should be noted that the length of a creative day is more than twenty-four hours. Most Bible scholars believe that a creative day refers to a period of a thousand years or more. The fact that the term "day" in the sense of the creative process did not necessarily refer to an earthly day can be discerned from Genesis 2:4, which uses the same Hebrew word for "day" but groups the six previous creative days into one day. So even though the "day and night" phrase is used in conjunction

---

[72] Gen. 1:1, 2, 9-10, 13 NIV

[73] Prov. 8:29

with the creative days to provide a human point of reference, a creative day was in actuality a reference to an event-filled period of time. Thus, multiple periods defined by independent events could be referred to individually as days, and these same events combined into one grand event could likewise be referred to as a day or period.

Most of us are more familiar with this usage than it appears. In fact, we have a similar usage of the word today. The phrases "back in the day" or "in my day" are typically used to refer to an undefined period of activity extending well beyond a single day. Therefore, it is reasonable to assert that these creative days were meant to represent periods of creation that extended beyond literal Earth days.

Regardless of the exact length of a creative day, the separation of the land and sea was neither constricted to a single day nor strung out across millions of years; this catastrophic event spanned a period of days to months. The instantaneous formation of mountains in the Genesis Hemispheric Impact Structure and the voluminous amounts of water pouring into the subsequently formed fractures dictate an extremely powerful catastrophic event that would have occurred at a relatively rapid rate.

Still, following the event there were undoubtedly major climatic changes due to the altered continental terrain, newly formed oceans establishing currents, and an atmosphere altered by the expulsion of water at impact. These factors would have left a large portion of the planet covered in ice. It would have required a substantial amount of time, perhaps thousands of years, for the oceans to warm and melt most or perhaps all of the planet's ice caps. A full or extensive thaw explains the remains of ancient vegetation and dinosaurs found below the ice caps, which currently envelop large regions such as Greenland and Antarctica. The thaw led to a unique time in Earth's history when virtually the entire planet offered climates far more suitable to life than what we find today.

According to the Biblical account, the planet existed in this paradisiacal state for quite some time. Then, again according to the account, just 1,656 years into man's existence[74]—about 4,400 years

---

[74] Based on Bible chronology found in Genesis chapters 1-7.

ago—a global flood occurred, wiping all life from the face of the earth except for eight human survivors and a menagerie of animals sealed within a large wooden ark.

The account relates:

> *The seventeenth day of the second month—on that day all the springs of the great deep burst forth, and the floodgates of the heavens were opened. And rain fell on the earth forty days and forty nights.*
>
> *The waters rose and increased greatly on the earth, and the ark floated on the surface of the water. They rose greatly on the earth, and all the high mountains under the entire heavens were covered. The waters rose and covered the mountains to a depth of more than fifteen cubits.*[75]

This passage claims that the "springs of the great deep," or the world's oceans, were the source of the sudden influx of water. That they "burst forth" signifies a violent event that rapidly expelled these waters from their deep oceanic basins on a global scale. If we are looking for a large source of water to drive a second COPHEE event, there is absolutely no better than the waters already filling the post-COPHEE I basins. And the violent event that would cause these basins to overflow? A terrestrial impact of similar magnitude to the Genesis impact.

Evidence of this impact appears to have been discovered in 2006 by a team of geological science researchers led by Ralph von Frese and Laramie Potts. The impact site was determined by the interpretation of gravity and subsurface radar maps, which revealed a 300-mile-diameter crater deep beneath Antarctica's icecap, known as the Wilkes Land crater. Included in the scientists' findings was the stunning hypothesis that the impact may have contributed to the fracture and separation of Antarctica from Australia due to its proximity to the portion of the Antarctic coastline that was once joined to the southern coast of Australia.

This makes the crater an all-too-perfect candidate for the COPHEE II impact. The span of seafloor crust that separates Australia and

---

[75] Gen. 7:11, 12, 18-20

Antarctica is an extension of the band of new seafloor crust lying along the Southeast Indian Ridge, which clearly signifies the two continents fractured and separated during the same period that the COPHEE II expansion occurred. In other words, these two continents, as well as V-ridges throughout the globe, were simultaneously separated by newer COPHEE II crust. This new crust lying between the two continents and extending out around the entire globe forms a path in the seafloor that we will refer to as the Noachian Expansion Belt (Fig. 135). This is the only major continental fracture and separation to have occurred after the COPHEE I event, and the fact that it occurred simultaneous to the second expansion links the Antarctic impact directly to the origins of COPHEE II.

Though this impact was less than half of the size of the Genesis impact, by the time it occurred, Earth had a vastly different crustal formation. This left the planet set for a unique effect. Earth's crust comprised a patchwork of two very distinct surfaces. Continental crust has an average thickness of 22 miles, while oceanic crust averages four miles in thickness and is far denser. More importantly, the oceanic basins back then were significantly deeper than they are today.

Following COPHEE I, the oceanic basins covered far less area on a planet that was much smaller than it is today. To contain the waters that fill today's oceans, which cover nearly twice the area, these basins would have had to exist at far greater depths. To illustrate, imagine if you had two similarly sized glasses filled with water and you removed one glass and moved its contents to the other glass. You would have to double the height or depth of the other glass in order to contain the extra volume. Similarly, if the oceanic basins covered nearly half of the area they cover today, they would have to have existed at a depth roughly twice the depth of today's oceans to retain the same waters.

While all of the world's oceans were smaller and deeper, none were deeper than the Pacific. When the initial breach in the Pacific occurred, the waters were entering the breach while the waters existed at their maximum depth, driving the water far deeper into Earth's surface than they would into the subsequent continental fractures that would form the other oceans. As the Atlantic, Indian, Arctic, and much of the Southern Ocean opened up, water levels pressed downward with

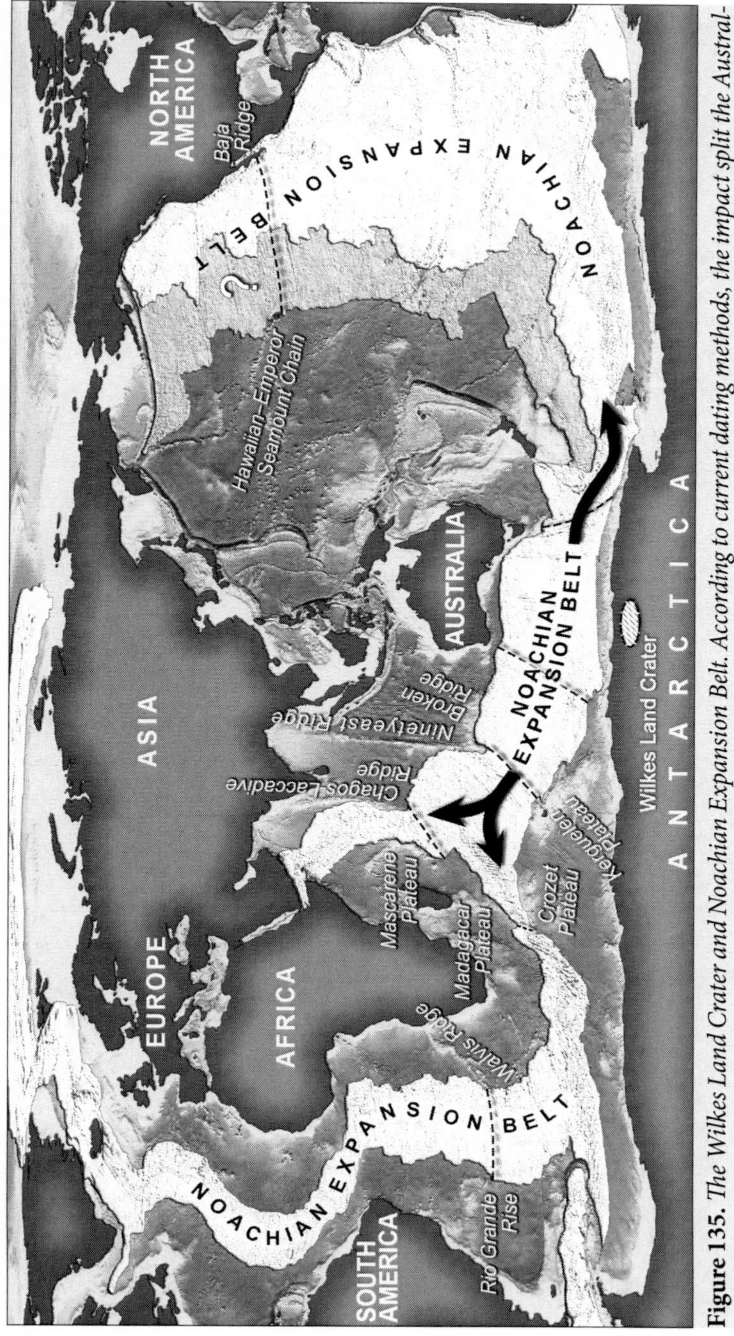

Figure 135. *The Wilkes Land Crater and Noachian Expansion Belt. According to current dating methods, the impact split the Austral-Antarctic continent, creating a span of seafloor between that extends out, separating seafloor ridges in the Atlantic and Indian Oceans. With evidence supporting a central divergent boundary in the North Pacific, which would contradict the current crustal age in the region, it is likely that the span between all ridges and the lands of Australia and Antarctica formed during the same expansion period.*

far less force, ripping the associated continents apart in a slightly more gradual fashion than the Pacific. This left these basins shallower and, aside from the ridges formed by the expansion wedges, with far less violent seafloor scarring.

It is within this setting that the COPHEE II impact object struck the center of the combined Austral-Antarctic continent. The object struck with such massive force that it drove the Austral-Antarctic crust downward until it reached its breaking point and the continent fractured in two. However, the impact also had an effect that extended immediately throughout the globe. Similar to how inward pressure applied to one point on an inflated ball applies outward pressure at all other points, the compression at the point of impact transitioned to outward force over the remainder of the globe.

Where the strength of a unified crust would likely contain the outward force and only experience slight temporary deformation during the impact, the patchwork surface of continental and seafloor crust of the post-COPHEE-I Earth would have reacted to the outward dispersal of energy in an entirely different manner. While the whole of the planet would expand outward to varying degrees, the lack of continuity in the crust would see the lower-lying oceanic basins pressed upward, emptying out a substantial amount of their contents.

Providing perspective, if the world were the size of a basketball, this crustal rise would occur within an outer layer little more than the thickness of a sheet of paper. On a planet exceeding 5,000 miles in diameter at the time of impact, Earth's relatively thin outer layer of crust would have hardly appeared altered along its perimeter when viewed from space. The surface of the earth, however, would have exhibited a drastic change. The entire planet, which had nearly equal parts land and water surface prior to impact, would have begun to transition to all blue as the oceans overran the land.

As the Biblical account of the global flood relates, the waters from "the great deep burst forth," propelled out of their basins by rising seafloors. The account states that these waters rose to eventually cover all land, with even the highest mountains under the heavens covered to a depth of 22.5 feet (15 cubits). With Mount Everest rising 29,000 feet above sea level, there is little chance that the waters could actually have risen to this height. The waters sat at this height prior to the Genesis event, when the Earth was much smaller and had a unified crust, but by the time the flood occurred Earth's surface area was roughly double the size and contained vast recessed basins. So, the waters rising to anywhere near such a height is unlikely.

This suggests that the scripture, though appearing to be a global statement, may have been conveying a more limited reference. The verse is likely a continuation of the preceding Biblical verses, providing an extended perspective of the floodwaters in relation to the ark. This would allow for an interpretation that "all the high mountains under the entire heavens," or as far as could be seen or engaged from the ark, were covered by water. Similar use of the expression "under the heavens" can be found in Deuteronomy 2:25 where God advises Moses before taking possession of Amorite land: "This very day I will begin to put the terror and fear of you on all the nations under heaven." Here again, the expression would not have applied to the entire globe, but to the Israelites and the nations with whom they interacted within a limited region.

Interpreting the writing in this more limited sense, which allows that the account's description of the floodwaters is presented in relation to the ark, also potentially provides insight into the significance of the statement that the waters "covered the mountains to a depth of more than fifteen cubits." Continuing on from the ark perspective, 22.5 feet, which is diminutive in relation to the earth but significant in relation to the ark, likely points to an important aspect of the ark. We are likely being informed that the waterline on the ark sat at 22.5 feet, or half of its 45-foot (30-cubit) height, and that the ark cleared Earth's surface, including the highest peaks in the region. This would therefore place these peaks at least 22.5 feet below the water's surface. It is possible that during the upheaval, the waters had risen from 10,000 to 20,000 feet above sea level. This would have potentially covered all of the mountains of the Middle East, including Mount Ararat, the stated final resting place of the ark.

It is also possible that pre-flood mountain ranges like the Himalayas, which were created by the Genesis impact, may have existed at a lower elevation prior to the global flood. Areas of imbalance in seafloor expansion following the Noachian impact may have driven an increase in the folding of continental crust where folding already existed. Atmospheric changes may have also added to the water level, but it is hard to imagine that this would have added any substantial depth. If all of the water in our current atmosphere fell to Earth, it would only amount to an inch of additional water.

Regardless, for all man and animals outside the ark to perish, as the Biblical account states, it is not necessary for all mountains throughout the globe to have been submerged. Any man or creature that somehow migrated to high peaks prior to the flood would have had to survive the rigors of living at high altitude. Thin air and lack of vegetation would have rendered these areas inhospitable, as they are today. Even man or beast existing at habitable high altitudes would not likely survive the cataclysmic effects brought on by the atmospheric and climatic changes during the forty days of intense continuous precipitation that followed the flood.

We are informed that following the waters bursting forth from their basins, "the floodgates of the heavens were opened. And rain fell on the earth forty days and forty nights." This would have wreaked havoc on any high elevations remaining above the waters. If the climate proved to be cold, as is typical at higher altitudes, forty days of constant torrential precipitation would have dropped an unimaginable amount of snow and ice. Even if somehow the ice were not present, saturated soils would have left any exposed peaks unstable. Liquefaction and mudslides would have occurred everywhere, leaving little likelihood of any creature surviving on a surface sitting above the floodwaters.

The catastrophic uplift of seafloor that emptied these waters out across the continents is also the more likely cause of a phenomenon discussed earlier in Chapter 10, the Benioff zones. While the earlier discussed folding dynamic is likely now in play to some extent, the initial fold is more likely the result of COPHEE II's seafloor uplift.

The West Pacific seafloor following the first COPHEE event sat more than 36,000 feet below sea level. This depth matches the lowest point on Earth today, the Mariana Trench. When the Austral-Antarctic impact forced the Pacific seafloor upward, excess crust at the edges of the Pacific basin pressed up from this depth, was forced outward, and folded in upon itself. The excessive depth of the pre-Noachian Pacific versus the other shallower oceans of the world coincides with the extensive Benioff zones and deeper trenches lying along the Pacific's perimeter.

The fact that the more pronounced folds occur along the western Pacific links back to the unique nature of the Pacific's origins. As discussed in the previous chapter, the shearing of Asia away from America left the two continents on slightly different planes, with Asia sitting lower than the Americas. This imbalance caused the seafloor to expand off to the west far more so than it did to the east. As a result, the pre-Noachian Pacific sat relatively shallow in the west and dropped down dramatically moving eastward. As the Pacific seafloor was thrust upward, the shallow seafloor crust along the Americas generated minimal folding. The much deeper eastern Pacific saw far more crust being thrust upward, resulting in the excessively deep folds and trenches lining the Asian Plate.

It remains significant that the Bible not only identifies the only viable source of the flood waters—waters already existing in the deep—but also notes that the waters did not recede immediately. It would only be after a period of 150 days that water levels would begin to drop.

> *The waters flooded the earth for a hundred and fifty days.*
>
> *The water receded steadily from the earth. At the end of the hundred and fifty days the water had gone down.*[76]

Earth's crust, which over time had become completely bonded and stable, having experienced little expansion since the original Genesis impact, was reluctant to begin the expansion process for a second time. Boundary ridges like the Hawaiian-Emperor seamount chain and the Ninetyeast Ridge, which had bonded significantly from upwelling lava, remained intact. However, a string of divergent boundaries, which exist as one extensive global structure, immediately reacted to the impact and slowly began to fracture around the globe. At the end of 150 days, these weakening crustal bonds experienced catastrophic failure, fracturing completely through. Similar to COPHEE I, the seafloor basins began

---

[76] Gen. 7:24, 8:2 NIV

expanding to accommodate all the waters that had overrun the continents.

The Biblical account of the flood also provides our first clue as to the rate of planetary expansion. "In the six hundredth year of Noah's life, on the seventeenth day of the second month,"[77] the Austral-Antarctic impact occurred and the waters inundated the entire planet. A year later, on "the twenty-seventh day of the second month the earth was completely dry."[78] That is a span of 370 days from global inundation until the water fully receded from the earth. With the expansion process initiated 150 days in, this allows that 220 days of expansion occurred—220 days from the time that the waters began to recede until they fully withdrew from Earth's surface.

V-shaped expansion ridges, which were all joined at their points along a global chain of divergent boundaries, were separated simultaneously throughout the globe as the divergent boundaries generated new crust for the expansion. No separation was greater in span than the Hawaiian-Emperor seamount chain from the Baja Peninsula. The crust lying between expanded 2,700 miles in 220 days, based on the Biblical account. This sets the rate of expansion in the Pacific at a phenomenal rate of over twelve miles per day. While scientists have recorded Earth's crust currently moving at the rate of less than one foot per year, during the expansion process the seafloor crust was spreading at a rate of nearly *one foot per second.*

Magnetic striping across this span between Hawaii and the Baja Peninsula records the energy expended during this rapid cataclysmic expansion, a record of Earth's unstable core. The Bible may actually address this sudden output of energy:

> *But God remembered Noah and all the wild animals and the livestock that were with him in the ark, and he sent a wind over the earth, and the waters receded.*[79]

The Hebrew word for wind, *ruah*, is also used to describe an

---

[77] Gen. 7:11

[78] Gen. 8:14

[79] Gen. 8:1

invisible force or energy. This verse could apply to the invisible energy generated as Earth's core became and remained highly active during the expansion process, only spinning down to its normal speed once there was a complete separation between land and sea.

Current science maintains that pole reversals have little effect on life, but the one source that chronicles COPHEE II provides details of incredible effects on life. According to the biblical account, prior to the global flood, the lifespans of man approached a thousand years—from Adam, who lived 930 years, to Noah, who entered the ark at 600 years and lived until the ripe old age of 950.

Following the flood, the ages of Noah's descendants decreased significantly. Noah's son Shem, nearly 100 years of age when he entered the ark, may have been affected by radiation or had his life cut short unnaturally after the flood, as he only lived to the age of 600. It would appear he did not escape genetic alteration, as the next three generations of descendants experienced post-flood lifespans consistently reduced by half. Removing the outliers of Enoch and Lamech, the average pre-flood lifespan was 930 years, while post-flood lifespans averaged 445 years over the next three generations.

A marked drop in lifespans occurs again with the generation that followed. Eber, the last of those to experience 400-year lifespans, gives birth to Peleg and lifespans again see a sudden drop by half.

> *Two sons were born to Eber: One was named Peleg, because in his time the earth was divided."*
>
> *When Peleg had lived 30 years, he became the father of Reu. And after he became the father of Reu, Peleg lived 209 years and had other sons and daughters.*[80]

From Peleg to Isaac, seven generations, the lifespans average 202 years, suggesting that these generations had also fallen victim to similar radiation exposure as was experienced at the time of the flood. And the Bible appears to link this genetic alteration to an expansion event, likely an extension of the COPHEE II event:

---

[80] Gen. 10:25; 11:18, 19

One was named Peleg, because in his time the earth was divided.[81]

Peleg translates to "division"; more specifically, it denotes a division by water. Further, the name Peleg has also been linked to the word "earthquake."[82] Taken altogether, we have all the signature elements of a COPHEE event: separation or division of land by water accompanied by tectonic activity and apparently, like COPHEE I, followed by reduced lifespans potentially tied to increased core activity and core-generated pole reversals.

Once again, we are faced with the challenge of identifying the source waters that initiated this event. However, the volume of water needed to cause this smaller event would be far less significant than that required to cover the earth at the time of the great flood. Unlike the flood, many survived this later event. This suggests a sudden influx of water driving a relatively minor COPHEE event, with the devastation mainly concentrated along low-lying coastal regions.

As might be expected following a global flood (and more significantly, a mass reconfiguration and enlargement of the oceanic basins), massive disruption of the world's oceanic currents would have caused the global climate to remain unstable for quite some time afterward. One of the results of this instability was the buildup of a cataclysmic time bomb, which sat looming over the unsuspecting populace for hundreds of years after the great flood.

There is a current consensus within the scientific community that ice ages had occurred in the ancient past, burying North America and northern Europe beneath a giant ice sheet. While the floodwaters were expanding the oceanic basins, climatic instability led to the formation of ice, which covered vast regions of the planet. After the COPHEE II expansion process came to a halt and the floodwaters had fully subsided from the continents, ice accumulation would continue, leading to a significant drop in sea level. Even with the drop in sea level, the ice that accumulated following the global flood comprised more water volume than the world's near-full oceanic basins could contain. These large ice sheets set the stage for what would happen next.

---

[81] Gen. 10:25

[82] Strong's Hebrew Lexicon, H6389

As the global climate moved toward stabilization following the flood, warming began to occur, releasing massive amounts of meltwater. The result was coastal flooding and further planetary expansion. The same secondary process undoubtedly occurred following the initial Genesis impact and COPHEE I event as the planet transitioned into a global tropical climate—but this time around, it impacted a populated planet.

Humanity had grown in numbers and established cities in coastal areas and low-lying inland plains, which were now being flooded as the oceans rose. At first, the waters rose above the exposed continental shelves, reclaiming the oceanic basins, which had seen a drop in water level due to the excessive ice accumulation. During this period, no expansion occurred.

As the icecaps continued to melt during Peleg's day, the meltwater began to overfill these basins, and seafloor expansion was once again initiated. The result being that the "earth was divided" further by water. As would be expected, the expansion caused shifting plates on a global scale, generating major earthquakes and in turn tsunamis.

This period of cataclysmic quakes and flooding could potentially tie the Peleg event to the same period surrounding the Atlantis event. Plato's account describes a period of significant flooding, but more importantly, it ties the flooding to major tectonic events.

> *There occurred violent earthquakes and floods; in a single day and night of misfortune all your warlike men in a body sank into the earth, and the island of Atlantis in like manner disappeared in the depths of the sea.*[83]

This falls perfectly in line with COPHEE activity. If sea levels were to rise rapidly by several feet, it would be enough to generate a global catastrophe. High-magnitude earthquakes would occur throughout the globe with tsunamis soon to follow. This resolves a major episode in the Atlantis tale, which describes flooding and earthquake activity occurring simultaneously on a grand scale in regions separated by thousands of miles and lying along nonadjacent seas. Ultimately described as a land "sunk by an earthquake," Atlantis experienced

---

[83] *Timaeus* 25c, d

tremors that saw its capital city disappear beneath the sea, while quakes in the Mediterranean likewise generated tsunamis, flooding coastal regions that included ancient Athens.

This cataclysmic end of an ice age may also have had another interesting consequence. If maps of ancient Antarctica were somehow found to be legitimate, they may point to a global climatic change that also saw an Antarctic continent transition rapidly from a land with little to no icecap to a continent completely enveloped in ice.

From the Noachian Flood to the days of Peleg and the demise of the Atlanteans, these ancient accounts describe incredible accounts of cataclysmic events on a global scale. Today's scientific community reduces or discounts these tales of rapid and cataclysmic global flooding, quakes, the division of land by water, and even global climate change to fit within a uniformitarian view of the world. This view maintains that major geological changes to Earth occurred over hundreds of millions of years, at the same slow incremental rate of change we witness today.

COPHEE carries the potential to break the hold uniformitarianism currently has on the scientific community and open the door for catastrophism. COPHEE also opens new possibilities in the way we interpret both historical and geological events of the past—and offers a glimpse at what might be in store for our planet going forward.

CHAPTER 15

MAPPING THE FUTURE

As we near the conclusion of this work, we are left with much to consider. What is clear is that maps of the past convey man's constantly changing worldview, just as maps and satellite images of the present continue to play a role in establishing our current worldview. These sources contain significant revelations about our world's history, geography, and even the very dynamics that drive our planet. It would be shortsighted to believe that no further revelations remain lurking within these sources both old and new, but discerning myth from fact is an important element of the analysis process. Of course, this applies to the findings discussed herein as well, which carry with them varying degrees of plausibility.

A question I have encountered with one of these findings is how Schöner could have done something as irresponsible as erroneously placing an ancient world map on the bottom of his world globe. This should elicit a follow-up question: If Schöner did not use a source map for this and his later 1524 globe, was it responsible cartographic practice for him to introduce freestyle designs for the southern continent? If he did not use source maps, how else do we reconcile the introduction of two very diverse designs? The fact that Schöner varies the two designs so radically—one a C-shaped ring encircling the south pole and the other composed of two joined solid masses overlying a good portion of the southern hemisphere—would suggest a far more irresponsible process of radical random design. On the other hand, the fact that both maps can be scaled using two distinct geographic points lends itself to the Piri Re'is' school of cartography of the same period, which he clearly states involves scaling maps of ancient landforms to new maps and globes.

At one point, I was contacted by an individual who shared that he too had determined that the unique lengthy waterway spanning one side of Schöner's 1515 southern continent was a depiction of the Nile River's source waters. He is a reputable map historian and has far more knowledge than I ever expect to have in the field. He explained that he had discerned the relationship of this waterway in relation to Schöner's African continent and not in relation to the continent upon which it was charted. He had written previously about this finding and believes that this depiction of the Nile is an obscure ancient view that locates the Nile's source waters on a separate land south of the African continent.

I asked him if he thought my theory was plausible. He replied in a very cordial but candid manner that it seemed far-fetched that Schöner would inadvertently place an ancient Roman world map on the bottom of his globe. While less than what I had hoped for, this response was essentially what I had expected. It is human nature to hold onto that which we already know or believe, and the more invested we are in that belief, the harder it becomes to devote time to considering or investigating contradictory theories that seem initially to carry little credibility. I cannot imagine that I would have reacted much differently if our roles were reversed.

Similarly, I came across a find I considered far-fetched when I began this trek fifteen years ago. It was only through a chance web search that I came across an image of an ancient map purported to depict Antarctica. The only reason I gave it any consideration at all was curiosity about how someone had arrived at such an implausible conclusion.

Though I was able to discern that the map, the Piri Re'is map, did not portray Antarctica, it began an unexpected journey of discovery. Do I believe that an ancient society charted a deglaciated Antarctica or that an ocean-going Atlantean civilization inhabited South America? I still maintain a healthy dose of skepticism due to some major obstacles that remain in the way. For example, when and how did an ancient map of Antarctica make its way into Medieval Europe and why is there a complete lack of archaeological evidence for an advanced seafaring people in South America? That said, I would no longer be surprised if evidence came to light that substantiated either find. I also would expect evidence to be found in the Parana Delta, South America, and the region around and including Carney and Siple Islands, Antarctica. Unfortunately, like any fringe belief, the potential evidence is nearly impossible to acquire. The high water table in the Parana Delta and thick layers of ice in Antarctica make archaeological excavations extremely difficult for both sites.

While these findings remain clearly on the fringe, the unexpected benefit of my investigation was the discovery of an ancient Roman world map once thought lost forever. I believe it will also go on to be a bit of a mind-bender, as I discovered it in the process of exploring Schöner's methodology of scaling an ancient map of Antarctica to his 1524 globe. Schöner's 1515 globe proves he portrayed a southern continent by scaling an ancient source map to new geographic discoveries. In 1524 he responded to Magellan's discovery of a land with a bay at the tip of South America and a pair of isolated islands in the Pacific using a large landform that looks very much like Antarctica; these are inaccurate portrayals of Magellan's two geographic discoveries that accurately depict Antarctica's Atka Bay and the islands of Siple and Carney precisely positioned relative to the landform. Assuming he again used an ancient source map to portray the southern continent, aside from Antarctica what other possible landform could the source map have originally portrayed?

And, of course, if not for my initial dismissive curiosity sending me on a trip into the fringe, I would not have stumbled upon discoveries that could change the way we look at the world and how we regard scientific consensus going forward.

The second half of this work brings forth a wealth of new findings pertaining to Earth dynamics that pose many challenges to the plate tectonics model. It all begins with a few small cracks in the foundation located in a place called Kamchatka. Recognizing Kamchatka as a fractured fragment of the Asian continent is the key to establishing this new dynamic. This is perhaps one of the more surprising oversights in the study of plate tectonics. Kamchatka exhibits very clear conformance with the adjacent Asian continent, with lone extensions off each coast that align perfectly to form an isthmus when the landforms are brought back together. This is a significant find that counters the belief that Kamchatka is the product of volcanic upwelling caused by subduction.

However, perhaps the larger oversight is overlooking the multitude of ductile fractures in the region. After all, fracturing is a basic element of plate tectonics, and ductile fracturing is as basic an element of fracture dynamics as brittle fracturing. This is a rather straightforward discovery that should find immediate acceptance, especially among those who have a background in the field of fracture mechanics, due to the overwhelming abundance of evidence.

The clear link between these continental ductile fractures and the ridges that extend out from them onto the adjacent seafloor crust dictates that continental crust is firmly affixed to the adjacent seafloor crust. When a fracture opens along a continental coastline, it breaks open the adjacent seafloor. This creates an expansion wedge, which is characterized by ridges defining the boundary between existing crust and the new crust filling the void between.

This link between continental cusps and seafloor ridges applies to the current alignment of the Hawaiian-Emperor seamount chain with a Kamchatka ductile cusp. This can only be explained by relatively minor seafloor folding having formed the deep seafloor trench lying between, which negates the possibility of seafloor subduction, a necessity of plate tectonics. Lacking subduction, Earth expansion is the only remaining possible dynamic driving plate movement and seafloor expansion.

Aside from the evidence provided by the Hawaiian-Emperor seamount chain alignment, perhaps the most spectacular and convincing finding in support of Earth expansion is the Genesis terrestrial impact crater and its associated hemispheric deformation. From the impact site on the Arctic seafloor, complete with central peak and surrounding terracing to the encircling hemispheric impact structure, there is little doubt that this is the site of the largest impact crater on the planet and the cause of the initial fracturing of Earth's crust.

Going forward, I do not believe anyone will be able to look at the northern hemispheric mountain ranges in quite the same way. The manner in which the mountains extend down along both sides of the Pacific and encircle the impact site in itself is significant. However, combined with the paralleling interior rifting forming the deepest inland bodies of water on Earth, it is difficult to imagine that anyone would continue to attribute this formation to a chance alignment of geographic features occurring through random plate movement and compression spanning billions of years. The fact that the compression zone and the land it encircles to the north comprise half the planet's continental mass makes it almost a certainty that we are looking at the aftermath of a hemispheric impact occurring on Earth when it existed at half its current size.

This brings us to the role played by water as it entered the breach in Earth's crust following the Genesis impact. It would be hard to deny that an ocean of water well over twice as deep as today's oceans would have an effect on the fracturing of a unified crust. Is it enough to drive plate movement? Is it plausible that the planet's oceans and oceanic basins could be directly linked to Earth expansion? Again, what *is* certain is that grand-scale subduction has not occurred, which leaves us with the lone option of Earth expansion. With the necessity for a powerful mechanism that not only drives the expansion process but is also linked to halting it, COPHEE, Catastrophic Ocean Planet Hydro-Equilibrial Expansion, offers an initial hypothesis that includes an immediately observable candidate for its trigger. Since water is the strongest candidate for initiating the expansion process, it makes sense to consider water when seeking an equilibrial off switch. It is hard to look past the fact that our vast oceanic basins exist filled to the brim and not beyond at a time when we see little to no expansion.

Accepting this new dynamic, we now recognize that the current fear of Earth being overrun by water on a long-term basis is an impossibility. Any major rise in sea level would initiate the expansion process, which would continue until sea level fell back to the brim of the expanded oceanic basins. This also establishes that major increases in water level have a direct link to seismic activity. It may even be the case that if the ice caps ceased melting, earthquakes would be nearly nonexistent. Prior to the great flood of COPHEE II, the planet likely attained this peaceful state, as the global climate had time following COPHEE I to warm to the point that even the Antarctic continent had achieved a tropical clime. Similarly, it is no coincidence that today we experience earthquakes throughout the globe at a time when the planet is in a state of warming following COPHEE II—warming that has led to the end of ice ages, shrinking glaciers, and dwindling icecaps.

Minus Earth's water, our planet would have been marked by the mountainous Genesis Hemispheric Impact Structure and a deep fracture in a unified crust lying between what is now North America and Asia. With no overlying waters to pry it open, what is now the Pacific basin would exist only as a large longitudinal canyon, and the planet would have remained at its original smaller size. There may actually exist another planet in our own solar system that serves as an example of the effects of a hemispheric impact on a planet devoid of a miles-deep layer of water.

There are many theories that attempt to explain the planet Mars' dramatically varied hemispheres—a crater-riddled raised terrain in the south and a smooth-surface hemisphere in the north. In the center of the raised surface of the planet's southern hemisphere lies a large void in the crust known as the Hellas impact crater. The Hellas impact crater, or *Hellas Planitia*, may sit amid a hemispheric impact structure similar to the Genesis Impact Crater with the debris ejected by the impact creating the crater-riddled terrain. Not only does the Hellas impact appear to have raised the southern hemisphere's surface, but like Earth, the impact may also be linked to the creation of a significant linear breach—the largest canyon in our solar system, *Valles Marineris*. With no sudden surge of overlying waters entering the freshly formed breach, as has happened on Earth, we see no expansion of the breach that could have potentially ripped Mar's unified crust into multiple continental plates.

Recognition of the Genesis Hemispheric Impact Structure also potentially resets the current geological time scale. The Appalachian and Himalaya Mountain ranges, currently believed to have been formed 450 million years ago and 45 million years respectively, were actually formed within minutes of each other. While the COPHEE events, which are tied to the accelerated spin of Earth's core, likely contribute to exaggerated results from radiometric dating, one must assume that improper techniques and misinterpretation of data must also drive these huge disparities.

Evidence of erroneous radiometric dating is not limited to the world's highest elevations; it also occurs beneath the world's oceans. When the Hawaiian Islands separated from the Baja Peninsula—along with all other V-shaped ridges throughout the globe—at the time of COPHEE II, the seafloor immediately adjacent to and between these two features was generated simultaneously. As the ridges separated during expansion, newer and newer spans of seafloor crust continually formed, centered between the two ridges. Yet radiometric dating of these regions conforms to the previously established view that the seafloor crust adjacent the Hawaiian Islands is older and gradually dates newer toward the Baja Peninsula. Therefore, where the crusts alongside these two ridges should date the same, radiometric dating has dated the crust near the Hawaiian Islands approximately 100 million years older than the crust lying alongside the Baja Peninsula.

These dating errors suggest that our current method of dating relies too heavily on a preconceived timeline for calibration or adjustment. Assuming, based on the above findings, that the radiometric dating process is laden with errors, we are left to question much or all of the currently established geological time scale. Not only does the current method of radiometric dating appear to be flawed, but with accelerated radioactive decay occurring during COPHEE events, it is rendered wholly unreliable and virtually useless.

The repercussions, if true, are staggering. Everything from Earth's true age to the first appearance of life on the planet comes into question. Of course, it also leaves the true age of man up in the air; humanity could suddenly be thousands rather than millions of years old. This could make the Bible's 6,000-year chronology of man as credible as any other posited timeline. This seems somewhat appropriate, as we have already seen that the Biblical accounts of creation and the Noachian Flood perfectly correspond to the two major catastrophic global events of COPHEE I and II.

Perhaps it is no coincidence then that this same ancient source may have accurately predicted global catastrophic events to come. According to Biblical prophecy, multiple dire and catastrophic events are foretold for the end times. Earthquakes on a global scale are foretold, along with the consequences of a global cataclysm.

> *There will be great earthquakes, famines and pestilences in various places.*[84]

We have already shown that the COPHEE dynamic dispels the possibility that the continents could become overrun and remain under water, but what would actually occur if Earth's glaciers and icecaps were to melt? The gradual melting of icecaps is likely responsible for most earthquakes that occur today, as the excess waters being added to the oceanic basins create an imbalance and place additional pressure on the basins and continental plates. The COPHEE dynamic also dictates that a more rapid melt would have a catastrophic effect on the globe.

If the Antarctic icecap were to experience a sudden, significant melt, the planet would experience earthquakes at a level unlike anything witnessed in recent history in both intensity and global scale. As the sudden surge of water overwhelms the oceanic basins and creates an imbalance, a COPHEE event would be triggered. The resulting plate movement would place an enormous amount of stress on major faults throughout the globe.

In North America, the San Andreas Fault would begin shifting. The state of California would be beset with quakes up and down the coast, wreaking havoc upon major cities including Los Angeles and San Francisco. Further north, the Cascadia subduction zone, which is actually a fold in the seafloor, would experience a shift within its fold, potentially recreating the catastrophic effects of the 1700 Cascadia earthquake. This could produce a tsunami of 80-100 feet in height, bringing mass devastation to the American Northwest.

Meanwhile the New Madrid Fault Line, already the site of the largest and most damaging quakes in American history, would likely witness quakes of equal or greater magnitude. These tremors would likely change the course of the Mississippi River and perhaps even open

---

[84] Luke 21:10, 11

up another chasm along it, forming another great lake. Major cities all along the Mississippi would be impacted, as well as most of the Midwest.

Across the Pacific, the Asian inland from China to Turkey—an area long prone to high-magnitude quakes—would experience unparalleled devastation. Quakes would occur along the Pacific Rim within the deep trenches formed by crustal folds. As a result, major island civilizations and cities all along the Asian coast would collapse due to high-magnitude quakes. The tsunamis that follow would add to the destruction, washing over the already decimated regions. Throughout the globe, coastal regions would be subjected to quakes and tsunamis mimicking the Atlantis cataclysm.

It is easy to picture the state of the world following these events. Those individuals that managed to survive would be subjected to disease and pestilence brought about by the inability to properly deal with the overwhelming amount of carnage left behind, a lack of electricity, and poor sanitation. All the while, little help would arrive from other communities, who would be themselves dealing with the aftermath. Major shipping lanes would experience disruption, and global devastation to railways and roads would completely isolate cities, resulting in a lack of basic supplies, including food and fresh water. A global famine would ensue. Humanity would face an extended struggle to reestablish itself, the result being conditions that parallel those of the biblical end times.

But that may not even be the worst of it. Based on the Biblical account of such events, humankind could potentially suffer an even more devastating blow—one that could ensure that civilization would never be able to return to its advanced technological state, and indeed threaten its very existence. As discussed earlier, a major portion of the COPHEE dynamic relates to the accelerated spin of Earth's core. As the oceanic basins fill beyond their limits and Earth expansion is initiated, the planet's core spins at an accelerated rate with the increased core activity rapidly flipping the magnetic poles, a consequence recorded in the seafloor's pattern of magnetic striping. The biblical chronology of man suggests that this period of excessive radioactivity results in a halving of man's lifespan. Imagine if the average age of man were to suddenly drop from eighty to forty years.

This would mean that a large portion of the population would not live to see their children into their teen years. If forty became the new eighty, many at forty would prematurely experience the effects of old age, with many becoming incapable of caring for their young children. If the educational institutions failed to reestablish themselves in the wake of the devastation, and in turn the medical field experienced a great decline, we could see average lifespans cut down to twenty to thirty years. It would be a crisis of epic proportions, and the human population would experience a steady decline.

Of course, many other things would have to occur to bring about these cataclysmic events, and the extent to which age would be affected, if at all, is impossible to predict. Yet ancient accounts like the fall of Atlantis and the Noachian Flood provide insight into the potential impact of the next COPHEE-related event, including the complete loss of civilizations and the survival and rise of new civilizations, should they follow. Determining the extent of fact within these accounts remains a challenge, but the legitimacy of COPHEE I and II may lead to research that sheds further light on these ancient accounts.

Separating myth from fact is a never-ending process in the ongoing search for answers to Earth's historical and geological past. As in times past, in the process of searching for answers, the creation of new myths is bound to occur. Some myths gain consensus, achieving global acceptance and establish worldview. I believe that plate tectonics will prove to be one of these established popular myths. If we examine the evidence offered herein and closely reevaluate the planet using the many resources available to us today, it will become clear that plate tectonics is a theory that has finally reached its end.

Throughout time, maps have conveyed the worldviews of the civilizations that charted them, exposing the blurred lines between myths and facts associated with lands traversed and seas boldly navigated. Ancient man charted maps of the world with depictions of grand mythical creatures populating unknown regions and even attributed volcanic activity to invisible entities dwelling beneath volcanoes. With the evolution of science and technology, modern man is charting the world at a level never seen before. However, while our explanations for today's mysteries are far more informed and complex, we still attribute ridges cut into the seafloor to powerful invisible forces that reside beneath volcanoes. Maps may now finally reveal the fallacy in these modern myths, leaving the door wide open for a new Earth dynamic.

COPHEE, anyone?

List of Figures

Index

N

O

P

Q

R